A Tutorial on Elliptic PDE Solvers and Their Parallelization

SOFTWARE • ENVIRONMENTS • TOOLS

The series includes handbooks and software guides as well as monographs
on practical implementation of computational methods, environments, and tools.
The focus is on making recent developments available in a practical format
to researchers and other users of these methods and tools.

Editor-in-Chief

Jack J. Dongarra
University of Tennessee and Oak Ridge National Laboratory

Editorial Board

James W. Demmel, University of California, Berkeley
Dennis Gannon, Indiana University
Eric Grosse, AT&T Bell Laboratories
Ken Kennedy, Rice University
Jorge J. Moré, Argonne National Laboratory

Software, Environments, and Tools

Craig C. Douglas
University of Kentucky
Lexington, Kentucky
and
Yale University
New Haven, Connecticut

Gundolf Haase
Johannes Kepler University
Linz, Austria

Ulrich Langer
Johannes Kepler University
Linz, Austria

A Tutorial on Elliptic PDE Solvers and Their Parallelization

Society for Industrial and Applied Mathematics
Philadelphia

Library of Congress Cataloging-in-Publication Data

Douglas, Craig C.
  A tutorial on elliptic PDE solvers and their parallelization / Craig C. Douglas, Gundolf Haase, Ulrich Langer.
    p.cm. — (Software, environments, tools)
    Includes bibliographical references and index.
    ISBN 0-89871-541-5 (pbk.)
Differential equations, Elliptic—Numerical solutions—Data processing. 2. Parallel
  algorithms. I. Haase, Gundolf, 1963- II. Langer, Ulrich, Dr. sc. nat. III. Title. IV. Series.

QA377.D688 2003
515'.353—dc21   2003042369

 is a registered trademark.

Contents

List of Figures

List of Algorithms

Abbreviations and Notation

Nobody can say what a variable is.
—Hermann Weyl (1885–1955)

ADI	Alternating Direction Implicit
a.e.	almost everywhere
AMG	Algebraic Multigrid
BC	Boundary Condition
BICGSTAB	BIConjugate Gradient STABilized (method, algorithm,...)
BPX	Bramble, Pasciak, and Xu Preconditioner
BVP	Boundary Value Problem
CFD	Computational Fluid Dynamics
CG	Conjugate Gradient (method, algorithm,...)
COMA	Cache Only Memory Access
CRS	Compressed Row Storage
DD	Domain Decomposition
DSM	Distributed Shared Memory
FDM	Finite Difference Method
FEM	Finite Element Method
FVM	Finite Volume Method
FFT	Fast Fourier Transformation
GMRES	Generalized Minimum RESidual (method, algorithm,...)
HPF	High Performance Fortran
IC	Incomplete Cholesky
ILU	Incomplete LU factorization
LU	LU factorization
MIC	Modified Incomplete Cholesky factorization
MILU	Modified Incomplete LU factorization
MIMD	Multiple Instructions on Multiple Data
MISD	Multiple Instructions on Single Data
MP	Material Property
MPI	Message-Passing Interface
NUMA	Nonuniform Memory Access
PCG	Preconditioned Conjugate Gradient
PDE	Partial Differential Equation

QMR	quasi-minimal residual
SIMD	Single Instruction on Multiple Data
SISD	Single Instruction on Single Data
SMP	Symmetric Multiprocessing
SOR	Successive Overrelaxation
SPMD	Single Program on Multiple Data
SSOR	Symmetric Successive Overrelaxation
SPD	Symmetric Positive Definite
UMA	Uniform Memory Accesss
$(\cdot, \cdot)$	inner (scalar) product in some Hilbert space V
$(\cdot, \cdot)_{1,\Omega}$	inner product in $H^1(\Omega)$: $(u, v)_{1,\Omega} = (u, v)_1 := \int_\Omega \left(\nabla^T u \nabla u + u^2 \right) dx$
$\| \cdot \|$	norm in some normed space V: $\| u \|^2 := (u, u)$ if V is a Hilbert space
$\| \cdot \|_{1,\Omega}$	norm in the function space H^1: $\| u \|^2_{1,\Omega} = \| u \|^2_1 := (u, u)_1$
$\langle F, \cdot \rangle$	continuous linear functional; see (4.3)
$a(\cdot, \cdot)$	bilinear form; see (4.3)
$(\underline{u}, \underline{v})$	Euclidian inner product in $\mathbb{R}^{N_h}$: $(\underline{u}, \underline{v}) = \sum_{i=1}^{N_h} u_i v_i \ \forall \underline{u}, \underline{v} \in \mathbb{R}^{N_h}$
$\| \underline{u} \|_B$	B-energy norm: $\| \underline{u} \|^2_B := (B\underline{u}, \underline{u})$, where B is an $N_h \times N_h$ SPD matrix
$\mathbb{R}$	real numbers
$\mathbb{R}^m$	m-dimensional Euclidean space
Ω	computational domain, $\Omega \subset \mathbb{R}^m$ ($m = 1, 2, 3$)
Ω_s	subdomain $\Omega_s \subset \Omega$
$\Gamma = \partial\Omega$	boundary of a domain Ω
Γ_1	Dirichlet boundary (section 4.1)
Γ_2	Neumann boundary (section 4.1)
Γ_3	Robin boundary (section 4.1)
$L_2(\Omega)$	function space: $L_2(\Omega) := \{v \mid \int_\Omega v(x)^2 dx < \infty\}$
$L_\infty(\Omega)$	function space: $L_\infty(\Omega) := \{v \mid \mathrm{esssup}_{x \in \Omega} |v(x)| < \infty\}$
$C(\Omega)$	space of continuous functions $f : \Omega \mapsto \mathbb{R}$
$C^m(\Omega)$	space of m times continuously differentiable functions $f : \Omega \mapsto \mathbb{R}$
X	general function space for classical solutions
$H^1(\Omega)$	Sobolev space $H^1(\Omega) := \left\{v(x) : \| v \|_{1,\Omega} < \infty\right\}$
$H^{\frac{1}{2}}(\Gamma_1)$	trace space $H^{\frac{1}{2}}(\Gamma_1) := \{g \mid \exists\, v \in H^1(\Omega) : v|_{\Gamma_1} = g\}$
V	Hilbert space
$V_0 \subset V$	subspace of V
$V_g \subset V$	linear manifold in V generated by a given $g \in V$ and V_0
V_0^*	dual space of all linear and continuous functionals on V_0
V_h, V_{0h}	finite dimensional subspaces: $V_h \subset V$, $V_{0h} \subset V_0$
V_{gh}	finite dimensional submanifold: $V_{gh} \subset V_g$
$\dim V_h$	dimension of space V_h (section 4.2.1)
$\overline{\omega}_h$	set of nodes or their indices from discretized $\overline{\Omega}$
γ_h	set of nodes or their indices from discretized Γ_1
ω_h	set of nodes or their indices from discretized $\overline{\Omega} \setminus \Gamma_1$
$\overline{N}_h$	number of nodes: $\overline{N}_h = |\overline{\omega}_h|$
N_h	number of nodes: $N_h = |\omega_h|$

u	classical solution $u \in X$ or weak solution $u \in V_g$
u_h	finite element (Galerkin) solution $u_h \in V_{gh}$
$\underline{u}_h$	discrete solution (vector of coefficients) $\underline{u}_h \in \mathbb{R}^{N_h}$
supp(φ)	supp$(\varphi) := \{x \in \Omega \mid \varphi(x) \neq 0\}$, support of $\varphi(x)$
$\mathcal{T}_h$	triangulation
R_h	set of finite elements
$\delta^{(r)}$	arbitrary triangular element
Δ	master (= reference) element
K_h, K	stiffness matrix
C_h, C	preconditioning matrix
$\kappa(K)$	condition number of a matrix K
δ_{ij}	Kronecker symbol, $\delta_{ij} = 1$ for $i = j$ and 0 otherwise
$O(N^p)$	some quantity that behaves proportional to the pth power of N for $N \to \infty$
$O(h^p)$	some quantity that behaves proportional to the pth power of h for $h \to 0$

Preface

Computing has become a third branch of research, joining the traditional practices of theorization and laboratory experimentation and verification. Due to the expense and complexity of actually performing experiments in many situations, simulation must be done first. In addition, computation in some fields leads to new directions in theory. The three practices have formed a symbiotic relationship that is now known as computational science. In the best situations, all three areas seamlessly interact.

The "computational" supplement is due to the wide use during the last decade of computational methods in almost all classical and new sciences, including life and business sciences.

Most of the models behind these computer simulations are based on partial differential equations (PDEs). The first step toward computer simulation consists of the discretization of the PDE model. This usually results in a very large-scale linear system of algebraic equations or even in a sequence of such systems. The fast solution of these systems of linear algebraic equations is crucial for the overall efficiency of the computer simulation. The growing complexity of the models increasingly requires the use of parallel computers. Besides expensive parallel computers, clusters of personal computers or workstations are now very popular as an inexpensive alternative parallel hardware configuration for the computer simulation of complex problems (and are even used as home parallel systems).

This tutorial serves as an introduction to the basic concepts of solving PDEs using parallel numerical methods. The ability to understand, develop, and implement parallel PDE solvers requires not only some basic knowledge about PDEs, discretization methods, and solution techniques, but also some knowledge about parallel computers, parallel programming, and the run-time behavior of parallel algorithms. Our tutorial provides this knowledge in just eight short chapters. We kept the examples simple so that the parallelization strategies are not dominated by technical details. The practical course for the tutorial can be downloaded from the internet (see Chapter 1 for the internet addresses).

This tutorial is intended for advanced undergraduate and graduate students in computational sciences and engineering. However, our book can be helpful to many people who use PDE-based parallel computer simulations in their professions. It is important to know at least something about the possible errors and bottlenecks in parallel scientific computing.

We are indebted to the reviewers, who contributed a lot to the improvement of our manuscript. In particular, we would like to thank Michael J. Holst and David E. Keyes for many helpful hints and words of advice. We would like to acknowledge the Austrian Science Fund for supporting our cooperation within the Special Research Program on "Numerical and Symbolic Scientific Computing" under the grant SFB F013 and the NSF for grants

DMS-9707040, CCR-9902022, CCR-9988165, and ACR-9721388.

Finally, we would like to thank our families, friends, and Gassners' Most & Jausn (`http://www.mostundjausn.at`), who put up with us, nourished us, and allowed us to write this book.

Greenwich, Connecticut (U.S.A.)   Craig C. Douglas
Linz (Austria)   Gundolf Haase and Ulrich Langer

September 2002

Chapter 1

Introduction

*A couple months in the laboratory can save a couple hours
in the library.*
—Frank H. Westheimer's Discovery

The computer simulation of physical phenomena and technical processes has become a powerful tool in developing new products and technologies. The computational supplement to many classical and new sciences, such as computational engineering, computational physics, computational chemistry, computational biology, computational medicine, and computational finance, is due to the wide use during the last decade of computational science methods in almost all sciences, including life and business sciences. Most of the models behind these computer simulations are based on partial differential equations (PDEs). After this modeling phase, which can be very complex and time consuming, the first step toward computer simulation consists of the discretization of the PDE model. This usually results in a large-scale linear system of algebraic equations or even in a sequence of such systems, e.g., in nonlinear and/or time-dependent problems. In the latter case, the fast solution of these systems of linear algebraic equations is crucial for the overall efficiency of the computer simulation. The growing complexity of the models increasingly requires the use of parallel computers. Clusters of workstations or even PCs are now very popular as the basic hardware configuration for the computer simulation of complex problems, and even as home parallel systems. Soon, with the forthcoming multiprocessor on a chip systems, multiprocessor laptops will make small-scale parallel processing commonplace, even on airplanes.

The correct handling of computer simulations requires interdisciplinary abilities and knowledge in different areas of applied mathematics and computer sciences and, of course, in the concrete field to which the application belongs. More precisely, besides some understanding of modeling, it requires at least basic knowledge of PDEs, discretization methods, numerical linear algebra, and, last but not least, computer science. The aim of this tutorial is to provide such basic knowledge not only to students in applied mathematics and computer science but also to students in various computational sciences and to people who use computational methods in their own research or applications.

Chapter 2 provides the first ideas about what goes on in a computer simulation based on PDEs. We look at the Poisson equation, which is certainly one of the most important PDEs, not only as the most widely used model problem for second-order elliptic PDEs, but also from a practical point of view. Fast Poisson solvers are needed in many applications, such as heat conduction (see also Example 4.3), electrical field computation (potential equation) [107], and pressure correction in computational fluid dynamics (CFD) simulation [36]. The simplest discretization method for the Poisson equation is certainly the classical finite difference method (FDM). Replacing the computational domain with some uniformly spaced grid, substituting second-order derivatives of the Laplace operator with second-order differences, and taking into account the boundary conditions (BCs), we arrive at a large-scale, sparse system of algebraic equations. As simple as the FDM is, many problems and difficulties arise in the case of more complicated computational domains, differential operators, and BCs. These difficulties can be overcome by the finite element method (FEM), which is discussed at length in Chapter 4. Nevertheless, due to its simple structure, the FDM has some advantages in some applications. The efficiency of the solution process is then mainly determined by the efficiency of the method for solving the corresponding linear system of algebraic equations, which is nothing more than the discrete representation of our PDE in the computer. In principle, there are two different classes of solvers: direct methods and iterative solvers. We will examine some representatives from each class in their sequential version in order to see their strengths and weaknesses with respect to our systems of finite difference equations. Finally, we move from a sequential to a parallel algorithm using elementary domain decomposition (DD) techniques.

Before using a parallel solver, the reader should be familiar with some basics of parallel computing. Chapter 3 starts with a rough but systematic view of the rapidly developing parallel computer hardware and how its characteristics have to be taken into account in program and algorithm design. The presentation focuses mainly on distributed memory computers as available in clusters of workstations and PCs. Only a few basic communication routines are introduced in a general way and they are sufficient to implement parallel solvers for PDEs. The interested reader can first try parallel programs just by using the standard message-passing interface (MPI) library and can start to create a personal library of routines necessary for the Exercises of the following chapters. The methodology of the parallel approach in this tutorial, together with the very portable MPI library, allows the development and running of these parallel programs on a simple PC or workstation because several parallel processes can run on one CPU. As soon as the parallel program is free of obvious errors, the codes can be recompiled and run on a real parallel computer, i.e., an expensive supercomputer or a much cheaper cluster of PCs (such as the Beowulf Project; see http://beowulf.org). If a programmer writes a parallel program, then it will not take long before someone else claims that a newer code is superior. Therefore, a few points on code efficiency in the context of parallel numerical software are given to assist the reader.

The FEM is nowadays certainly the most powerful discretization technique on Earth for elliptic boundary value problems (BVPs). Due to its great flexibility the FEM is more widely used in applications than the FDM considered in our introductory chapter for the Poisson equation. Chapter 4 gives a brief introduction to the basic mathematical knowledge that is necessary for the implementation of a simple finite element code from a mathematical viewpoint. It is not so important to know a lot about Sobolev spaces, but it is important to be able to derive the variational, or weak, formulation of the BVP that we are going to solve from its classical statement. The basic mathematical tool for this is the formula of integration by parts. In contrast to the FDM, where we derive the finite difference equations directly from the PDE by replacing the derivatives with differences, the FEM starts with the variational formulation of the BVP and looks for an approximation to its exact solution in the form of some linear combination of a finite number of basis functions with local support. The second part of Chapter 4 provides a precise description of the procedure for deriving the finite element equations in the case of linear triangular finite elements. This procedure can easily be generalized to other types of finite elements. After reading this part, the reader should be able to write a first finite element code that generates the finite element equations approximating the BVP. The final and usually the most time-consuming step in finite element analysis consists of the solution of the finite element equations, which is nothing more than a large, sparse system of algebraic equations. The parallelization of the solver of the finite element equations is the most important part in the parallelization of the entire finite element code. The parallelization of the other parts of a finite element code is more or less straightforward. The solution methods and their parallelization are mainly discussed in Chapters 5 and 6. We will see in Chapter 2 that iterative solvers are preferred for really large scale systems, especially if we are thinking of their parallel solution. That is why we focus our attention on iterative solvers. Some theoretical aspects that are directly connected with the iterative solution of the finite element equations are considered at the end of Chapter 4. There is also a subsection (section 4.2.3) that discusses briefly the analysis of the FEM. This subsection can be skipped by the reader who is only interested in the practical aspects of the FEM.

Before the parallel numerical algorithms for solving the discrete problems, Chapter 5 investigates basic numerical features needed in all of the appropriate solution algorithms with respect to their adaptation to parallel computers. Here we concentrate on a data decomposition that is naturally based on the FEM discretization. We see that numerical primitives such as inner product and matrix-vector multiplication are amazingly easy to parallelize. The Exercises refer to Chapter 3 and they will lead the reader directly to a personal parallel implementation of data manipulations on parallel computers.

Conventional classical direct and iterative solution methods for solving large systems of algebraic equations are analyzed for their parallelization properties in Chapter 6. Here the detailed investigations of basic numerical routines from the previous chapter simplify the parallelization significantly, especially for iterative solvers. Again, the reader can implement a personal parallel solver guided by the Exercises.

Classical solution methods suffer from the fact that the solution time increases much faster than the number of unknowns in the system of equations; e.g., it takes the solver 100–1000 times as long when the number of unknowns is increased by a factor of 10. Chapter 7 introduces briefly a multigrid solver that is an optimal solver in the sense that it is 10 times as expensive with respect to both memory requirements and solution time for 10 times as many unknowns. Thanks to the previous chapters, the analysis of the parallelization properties is rather simple for this multigrid algorithm. If the reader has followed the practical course up to this stage, then it will be easy to implement a first parallel multigrid algorithm.

Chapter 8 addresses some problems that are not discussed in the book but that are closely related to the topics of this tutorial. The references given there should help the reader to generalize the parallelization techniques presented in this tutorial to other classes of PDE problems and applications. We also refer the reader to [45], [104], and [53] for further studies in parallel scientific computing.

A list of abbreviations and notation is provided on pages xiii–xiv. Finally, the Appendix provides a guide to the internet.

The practical course for the tutorial can be downloaded from

`http://www.numa.uni-linz.ac.at/books#Douglas-Haase-Langer`

or

`http://www.mgnet.org/mgnet-books.html#Douglas-Haase-Langer.`

Theory cannot replace practice, but theory can greatly enhance the understanding of what is going on in the computations. Do not believe the computational results blindly without knowing something about possible errors, such as modeling errors, discretization errors, round-off errors, iteration errors, and, last but not least, programming errors.

Chapter 2

A Simple Example

> *The things of this world cannot be made known without a knowledge of mathematics.*
> —Roger Bacon (1214–1294)

2.1 The Poisson equation and its finite difference discretization

Let us start with a simple, but at the same time very important, example, namely, the Dirichlet boundary value problem (BVP) for the Poisson equation in the rectangular domain $\Omega = (0, 2) \times (0, 1)$ with the boundary $\Gamma = \partial\Omega$: Given some real function f in Ω and some real function g on Γ, find a real function $u : \overline{\Omega} \to \mathbb{R}$ defined on $\overline{\Omega} := \Omega \cup \Gamma$ such that

$$
\begin{aligned}
-\Delta u(x, y) := -\frac{\partial^2 u}{\partial x^2}(x, y) - \frac{\partial^2 u}{\partial y^2}(x, y) &= f(x, y) \quad \forall (x, y) \in \Omega, \\
u(x, y) &= g(x, y) \quad \forall (x, y) \in \Gamma.
\end{aligned}
\tag{2.1}
$$

The given functions as well as the solution are supposed to be sufficiently smooth. The differential operator Δ is called the Laplace operator. For simplicity, we consider only homogeneous Dirichlet boundary conditions (BCs); i.e., $g = 0$ on Γ.

The Poisson equation (2.1) is certainly the most prominent representative of second-order elliptic partial differential equations (PDEs), not only from a practical point of view, but also as the most frequently used model problem for testing numerical algorithms. Indeed, the Poisson equation is the model problem for elliptic PDEs, much like the heat and wave equations are for parabolic and hyperbolic PDEs.

The Poisson equation can be solved analytically in special cases, such as rectangular domains with just the right BCs, e.g., with homogeneous Dirichlet BCs, as imposed above. Due to the simple structure of the differential operator and the simple domain in (2.1), a considerable body of analysis is known that can be used to derive or verify solution methods for Poisson's equation and other more complicated equations.

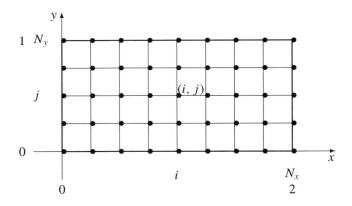

Figure 2.1. *Rectangular domain with equidistant grid points.*

If we split the intervals in both the x and y directions into N_x and N_y subintervals of length h each (i.e., $N_x = 2N_y$), then we obtain a grid (or mesh) of nodes like the one presented in Fig. 2.1. We define the set of subscripts for all nodes by $\overline{\omega}_h := \{(i, j) : i = \overline{0, N_x}, \ j = \overline{0, N_y}\}$, the set of subscripts belonging to interior nodes by $\omega_h := \{(i, j) : i = \overline{1, N_x - 1}, \ j = \overline{1, N_y - 1}\}$, and the corresponding set of boundary nodes by $\gamma_h := \overline{\omega}_h \setminus \omega_h$. Furthermore, we set $f_{i,j} = f(x_i, y_j)$ and we denote the approximate values of the solution $u(x_i, y_j)$ of (2.1) at the grid points $(x_i, y_j) := (ih, jh)$ by the values $u_{i,j} = u_h(x_i, y_j)$ of some grid function $u_h : \overline{\omega}_h \to \mathbb{R}$. Here and in the following we associate the set of indices with the corresponding set of grid points; i.e., $\overline{\omega}_h \ni (i, j) \leftrightarrow (x_i, y_j) \in \overline{\Omega}_h$. Replacing both second derivatives in (2.1) with second-order finite differences at the grid points:

$$\frac{\partial^2 u}{\partial x^2}(x_i, y_j) \approx \frac{1}{h^2}\left(u_{i-1,j} - 2u_{i,j} + u_{i+1,j}\right) \qquad \text{and} \tag{2.2}$$

$$\frac{\partial^2 u}{\partial y^2}(x_i, y_j) \approx \frac{1}{h^2}\left(u_{i,j-1} - 2u_{i,j} + u_{i,j+1}\right), \tag{2.3}$$

we immediately arrive at the following five-point stencil finite difference scheme that represents the discrete approximation to (2.1) on the grid $\overline{\omega}_h$: Find the values $u_{i,j}$ of the grid function $u_h : \overline{\omega}_h \to \mathbb{R}$ at all grid points (x_i, y_j), with $(i, j) \in \overline{\omega}_h$, such that

$$\boxed{\begin{aligned} \frac{1}{h^2}\left(-u_{i,j-1} - u_{i-1,j} + 4u_{i,j} - u_{i+1,j} - u_{i,j+1}\right) &= f_{i,j} \quad \forall (i, j) \in \omega_h, \\ u_{i,j} &= 0 \quad \forall (i, j) \in \gamma_h. \end{aligned}} \tag{2.4}$$

Arranging the interior (unknown) values $u_{i,j}$ of the grid function u_h in a proper way in some vector $\underline{u}_h$, e.g., along the vertical (or horizontal) grid lines, and taking into account the BCs on γ_h, we observe that the finite difference scheme (2.4) is equivalent to the following system of linear algebraic equations: Find $\underline{u}_h \in \mathbb{R}^N$, $N = (N_x - 1) \cdot (N_y - 1)$, such that

$$\boxed{K_h \underline{u}_h = \underline{f}_h,} \tag{2.5}$$

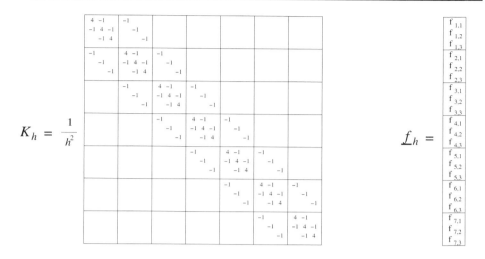

$$K_h = \frac{1}{h^2}$$

$$\underline{f}_h =$$

Figure 2.2. *Structure of matrix and load vector.*

where the $N \times N$ system matrix K_h and the right-hand side vector $\underline{f}_h \in \mathbb{R}^N$ can be rewritten from (2.4) in an explicit form. This is illustrated in Fig. 2.2.

The (band as well as profile) structure of the matrix K_h heavily depends on the arrangement of the unknowns, i.e., on the numbering of the grid points. In our example, we prefer the numbering along the vertical grid lines because exactly this numbering gives us the smallest bandwidth in the matrix. Keeping the bandwidth as small as possible is very important for the efficiency of direct solvers.

Before reviewing some solution methods, we first look at further algebraic and analytic properties of the system matrix K_h that may have some impact on the efficiency of solvers, especially for large-scale systems. The finer the discretization, the larger the system is. More precisely, the dimension N of the system grows like $O(h^{-m})$, where m is the dimension of the computational domain Ω; i.e., $m = 2$ for our model problem (2.1). Fortunately, the matrix K_h is sparse. A *sparse matrix* is one with only a few nonzero elements per matrix row and column, independent of its dimension. Our matrix K_h in (2.5) has at most five nonzero entries per row and per column, independent of the fineness of the discretization. This property is certainly the most important one with respect to the efficiency of iterative as well as direct solvers.

A smart discretization technique should preserve the inherent properties of the differential operator involved in the BVP. In our case, the matrix K_h is symmetric ($K_h = K_h^T$) and positive definite (($K_h \underline{u}_h, \underline{u}_h) > 0 \;\; \forall \, \underline{u}_h \neq 0$). These properties result from the symmetry (formal self-adjointness) and uniform ellipticity of the Laplace operator. Symmetric positive definite (SPD) matrices are regular. *Regular matrices* are invertible. Thus, our system of finite difference equations (2.4) has a unique solution.

Unfortunately, the matrix K_h is badly conditioned. The spectral condition number $\kappa(K_h) := \lambda_{max}(K_h)/\lambda_{min}(K_h)$ defined by the ratio of the maximal eigenvalue $\lambda_{max}(K_h)$ and the minimal eigenvalue $\lambda_{min}(K_h)$ of the matrix K_h behaves like $O(h^{-2})$ if h tends to

0. That behavior affects the convergence rate of all classical iterative methods and it can deteriorate the accuracy of the solution obtained by some direct method due to accumulated round-off errors, especially on fine grids. These properties are not only typical features of matrices arising from the finite difference discretization, but also characteristic of matrices arising from the finite element discretization discussed in Chapter 4.

2.2   Sequential solving

There exist two vastly different approaches for solving (2.5): direct methods and iterative methods. For notational purposes we consider an additive splitting of the matrix $K = E + D + F$ into its strictly lower part E, its diagonal part $D = \text{diag}(K)$, and its strictly upper part F. Here and in the following we omit the subscript h.

2.2.1   Direct methods

Direct methods produce the exact solution of systems of algebraic equations just like (2.5) in a finite number of arithmetical operations, provided that exact arithmetic without round-off errors is used.

The classical direct method is the *Gauss elimination* process, which successively replaces all entries in E with 0 and updates in each step the remaining part of the matrix and the right-hand side $\underline{f}$ [46]. This step is called the elimination step. The resulting transformed system of equations is triangular:

$$\left(\widetilde{D} + \widetilde{F}\right)\underline{u} \;=\; \widetilde{\underline{f}}, \tag{2.6}$$

and can easily be solved by replacing the unknowns from the bottom to the top. This step is called backward substitution.

The Gauss elimination produces in fact the so-called LU factorization $K = LU$ of the matrix K into a lower triangular matrix L, where all main diagonal entries are equal to 1, and an upper triangular matrix U. After this factorization step the solution of our system (2.5) reduces to the following forward and backward substitution steps for the triangular systems:

$$L\underline{x} \;=\; \underline{f} \qquad \text{and} \qquad U\underline{u} \;=\; \underline{x}, \tag{2.7}$$

respectively. If the matrix K is symmetric and regular, then the LU factorization can be put into the form $K = LDL^T$, extracting some (block) diagonal matrix D. In the SPD case, the *Cholesky factorization* $K = U^T U$ is quite popular [46, 80]. There exist several modifications of the Gaussian elimination method and the factorization techniques, but all of them use basically the same principles [46, 80, 94].

Each of these direct methods suffers from so-called fill-in during the elimination, or factorization, step. After the factorization the triangular factors become much denser than the original matrix. Nonzero entries appear at places within the band or profile of the triangular factors where zero entries were found in the original matrix. The sparsity of the original matrix gets lost. The fill-in phenomenon causes a superlinear growth of complexity. More precisely, the number of arithmetical operations and the storage requirements grow like $O(h^{-3m+2})$ and $O(h^{-2m+1})$, respectively, as the discretization parameter h tends to 0.

In addition to this, one must be aware of the loss of about $\log(\kappa(K))$ valid digits in the solution due to round-off errors.

The cost of computing the LU factorization and the substitutions in (2.7) is always less than the cost of computing $K^{-1}\widetilde{f}$ [65]. Therefore, K^{-1} should never be explicitly computed.

2.2.2 Iterative methods

Iterative methods produce a sequence $\{\underline{u}^k\}$ of iterates that should converge to the exact solution $\underline{u}$ of our system (2.5) for an arbitrarily chosen initial guess $\underline{u}^0 \in \mathbb{R}^N$ as k tends to infinity, where the superscript k is the iteration index.

In this introductory section to iterative methods we only consider some classical iterative methods as special cases of stationary iterative methods of the form

$$C\frac{\underline{u}^{k+1} - \underline{u}^k}{\tau} + K\underline{u}^k = \underline{f}, \quad k = 0, 1, \ldots, \tag{2.8}$$

where τ is some properly chosen iteration (relaxation) parameter and C is a nonsingular matrix (sometimes called preconditioner) that can be inverted easily and hopefully improves the convergence rate. If K and C are SPD, then the iteration process (2.8) converges for an arbitrarily chosen initial guess $\underline{u}^0 \in \mathbb{R}^N$ provided that τ was picked up from the interval $(0, 2/\lambda_{max}(C^{-1}K))$, where $\lambda_{max}(C^{-1}K)$ denotes the maximal eigenvalue of the matrix $C^{-1}K$ (see also section 4.2.4).

Setting $C := I$ results in the classical *Richardson iteration*

$$\underline{u}^{k+1} := \underline{u}^k + \tau\left(\underline{f} - K\underline{u}^k\right). \tag{2.9}$$

An improvement consists of the choice $C := D$ and leads to the *ω-Jacobi iteration* ($\tau = \omega$)

$$\underline{u}^{k+1} := \underline{u}^k + \omega D^{-1}\left(\underline{f} - K\underline{u}^k\right), \tag{2.10}$$

which is called the Jacobi iteration for $\omega = 1$.

Choosing $C = D + E$ (the lower triangular part of matrix K, which can be easily inverted by forward substitution) and $\tau = 1$ yields the *forward Gauss–Seidel iteration*

$$(D + E)\underline{u}^{k+1} = \underline{f} - F\underline{u}^k. \tag{2.11}$$

Replacing $C = D + E$ with $C = D + F$ (i.e., changing the places of E and F in (2.11)) gives the *backward Gauss–Seidel iteration*.

The slightly different choice $C = D + \omega E$ and $\tau = \omega$ results in the *successive overrelaxation (SOR) iteration*

$$(D + \omega E)\underline{u}^{k+1} = (1 - \omega)D\underline{u}^k + \omega\left(\underline{f} - F\underline{u}^k\right). \tag{2.12}$$

Changing again the places of E and F in (2.12) gives the backward version of the SOR iteration in analogy to the backward version of the Gauss–Seidel iteration. The forward

SOR step (2.12) followed by a backward SOR step gives the so-called *symmetric successive overrelaxation (SSOR) iteration.* The SSOR iteration corresponds to our basic iteration scheme (2.8) with the SSOR preconditioner

$$C = (D + \omega E)D^{-1}(D + \omega F) \tag{2.13}$$

and the relaxation parameter $\tau = \omega(2 - \omega)$. The correct choice of the relaxation parameter $\omega \in (0, 2)$ is certainly one crucial point for obtaining reasonable convergence rates. The interested reader can find more information about these basic iterative methods and, especially, about the correct choice of the relaxation parameters, e.g., in [92, pp. 95–116].

The Laplace operator in (2.1) is the sum of two one-dimensional (1D) differential operators. Both have been discretized separately by a three-point stencil resulting in regular $N \times N$ matrices K_x and K_y. Therefore, we can express the matrix in (2.5) as the sum $K = K_x + K_y$. Each row and column in the discretization (Fig. 2.1) corresponds to one block in the block diagonal matrices K_x and K_y. All these blocks are tridiagonal after a temporary and local reordering of the unknowns. We rewrite (2.5) as

$$\left(K_x + K_y\right)\underline{u} = \left(K_x + \rho I\right)\underline{u} + \left(K_y - \rho I\right)\underline{u} = \underline{f},$$

giving us the equation

$$(K_x + \rho I)\underline{u} = \underline{f} - \left(K_y - \rho I\right)\underline{u},$$

which can easily be converted into the iteration form

$$(K_x + \rho I)\underline{u}^{l+1} = \underline{f} - \left(K_x + K_y\right)\underline{u}^l + (K_x + \rho I)\underline{u}^l.$$

This results in an iteration in the x direction that fits into scheme (2.8):

$$(K_x + \rho I)\left(\underline{u}^{l+1} - \underline{u}^l\right) = \underline{f} - K\underline{u}^l.$$

A similar iteration can be derived for the y direction. Combining both iterations as half-steps in one iteration, just as we combined the forward and backward SOR iterations, defines the *alternating direction implicit iterative method,* known as the *ADI method*:

$$(K_x + \rho I)\underline{u}^{k+1/2} = \underline{f} - \left(K_y - \rho I\right)\underline{u}^k,$$
$$\left(K_y + \rho I\right)\underline{u}^{k+1} = \underline{f} - (K_x - \rho I)\underline{u}^{k+1/2}. \tag{2.14}$$

We refer to section 6.2.3 and to [92, pp. 116–118] for more information about the ADI methods.

The classical iteration methods suffer from the bad conditioning of the matrices arising from finite difference, or finite element, discretization. An appropriate preconditioning and the acceleration by Krylov space methods can be very helpful. On the other hand, the classical (eventually, properly damped) iteration methods usually have good smoothing properties; i.e., the high frequencies in a Fourier decomposition of the iteration error are damped out much faster than the low frequencies. This smoothing property combined with the coarse grid approximation of the smooth parts leads us directly to multigrid methods, which are discussed in Chapter 7. The parallelization of the classical and the advanced

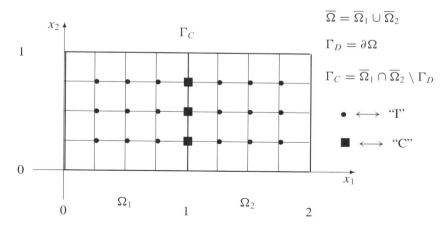

Figure 2.3. *Two subdomains with five-point stencil discretization.*

iteration methods (e.g., the multigrid methods, where we need the classical iteration methods as smoothers, and the direct methods as coarse grid solvers) is certainly the most important step toward the efficient solution of really large scale problems. Therefore, in the next section, we briefly consider the domain decomposition (DD) method, which is nowadays the basic tool for constructing parallel solution methods.

2.3 Parallel solving by means of DD

The simplest nonoverlapping DD for our rectangle consists of a decomposition of $\overline{\Omega}$ into the two unit squares $\overline{\Omega}_1 = [0, 1] \times [0, 1]$ and $\overline{\Omega}_2 = [1, 2] \times [0, 1]$; see Fig. 2.3. The interface between the two subdomains is Γ_C. This implies two classes of nodes. The coupling nodes on the interface are denoted by subscript "C" and the interior nodes by subscript "I." These classes of nodes yield the block structure (2.15) in system (2.5):

$$\begin{pmatrix} K_C & K_{CI,1} & K_{CI,2} \\ K_{IC,1} & K_{I,1} & 0 \\ K_{IC,2} & 0 & K_{I,2} \end{pmatrix} \begin{pmatrix} \underline{u}_C \\ \underline{u}_{I,1} \\ \underline{u}_{I,2} \end{pmatrix} = \begin{pmatrix} \underline{f}_C \\ \underline{f}_{I,1} \\ \underline{f}_{I,2} \end{pmatrix}. \qquad (2.15)$$

There are no matrix entries between interior nodes of different subdomains because of the locality of the five-point stencil used in our finite difference discretization.

A block Gaussian elimination of the upper-right entries of the matrix (which corresponds to the elimination of the interior unknowns) results in the block triangular system of equations

$$\begin{pmatrix} S_C & 0 \\ K_{IC} & K_I \end{pmatrix} \begin{pmatrix} \underline{u}_C \\ \underline{u}_I \end{pmatrix} = \begin{pmatrix} \underline{g}_C \\ \underline{f}_I \end{pmatrix}, \qquad (2.16)$$

with the Schur complement

$$S_C = K_C - K_{CI}K_I^{-1}K_{IC} \tag{2.17}$$
$$= \left(K_{C,1} - K_{CI,1}K_{I,1}^{-1}K_{IC,1}\right) + \left(K_{C,2} - K_{CI,2}K_{I,2}^{-1}K_{IC,2}\right)$$

and a modified right-hand side

$$\underline{g}_C = \underline{f}_C - K_{CI}K_I^{-1}\underline{f}_I \tag{2.18}$$
$$= \left(\underline{f}_{C,1} - K_{CI,1}K_{I,1}^{-1}\underline{f}_{I,1}\right) + \left(\underline{f}_{C,2} - K_{CI,2}K_{I,2}^{-1}\underline{f}_{I,2}\right).$$

Therein, $K_C = K_{C,1} + K_{C,2}$, $\underline{f}_C = \underline{f}_{C,1} + \underline{f}_{C,2}$, $K_{IC} = \begin{pmatrix} K_{IC,1} \\ K_{IC,2} \end{pmatrix}$, and $K_I =$ blockdiag$\{K_{I,1}, K_{I,2}\}$. The local Schur complements $K_{C,s} - K_{CI,s}K_{I,s}^{-1}K_{IC,s}$ are completely dense in general.

I) $\underline{g}_C := \underline{f}_C$
     for $s := 1$ to P do in parallel
          $\underline{g}_C := \underline{g}_C - K_{CI,s} \cdot K_{I,s}^{-1} \cdot \underline{f}_{I,s}$
     end

II) Solve $S_C\underline{u}_C = \underline{g}_C$

III) for $s := 1$ to P do in parallel
          $\underline{u}_{I,s} := K_{I,s}^{-1} \cdot \left(\underline{f}_{I,s} - K_{IC,s} \cdot \underline{u}_C\right)$
     end

Algorithm 2.1: A simple parallel DD solver ($P = 2$ for our example).

This procedure of eliminating the interior unknowns can obviously be extended to the more general case of decomposition into P subdomains. This generalization is depicted in Algorithm 2.1 ($P = 2$ for our example of the decomposition of $\overline{\Omega}$ into the two subdomains $\overline{\Omega}_1$ and $\overline{\Omega}_2$). The main problem in Algorithm 2.1 consists of forming and solving the Schur complement system in step II. In our simple example, we can calculate S_C explicitly and solve it directly. This is exactly the same approach that was used in the classical finite element substructuring technique. In general, the forming and the direct solution of the Schur complement system in step II of Algorithm 2.1 is too expensive. The iterative solution of the Schur complement system (sometimes also called the iterative substructuring method) requires only the matrix-by-vector multiplication $S_C \cdot \underline{u}_C^k$ and eventually a preconditioning operation $\underline{w}_C^k = C_C^{-1}\underline{d}_C^k$ as basic operations, where $\underline{d}_C^k = \underline{g}_C - S_C \cdot \underline{u}_C^k$ denotes the defect after k iterations (see section 2.2.2). The matrix-by-vector multiplication $S_C \cdot \underline{u}_C^k$ involves the solution of small systems with the matrices $K_{I,s}$ that can be carried out completely in parallel. The construction of a really good Schur complement preconditioner is a challenging task. Nowadays, optimal, or at least almost optimal, Schur complement preconditioners are available [97]. In our simple example, we can use a preconditioner proposed by M. Dryja [33]. Dryja's preconditioner replaces the Schur complement with the square root of the discretized 1D (one-dimensional) Laplacian K_y along the interface (see section 2.2.2) using the fact that K_y has the discrete eigenvectors $\underline{\mu}_l = [\mu_l(i)]_{i=0}^{N_y} = [\sqrt{2}\sin(l\pi ih)]_{i=0}^{N_y}$ and eigenvalues $\lambda_l(K_y) = \frac{4}{h^2}\sin^2(l\pi h/2)$, with $l = 1, 2, \ldots, N_y - 1$. This allows us to express the defect $\underline{d}_C$ and the solution $\underline{w}_C$ of the preconditioning equation (we omit the iteration index k for simplicity) as linear combinations of the eigenvectors $\underline{\mu}_l$:

$$\underline{d}_C = \sum_{l=1}^{N_y-1} \gamma_l \cdot \underline{\mu}_l \quad \text{and} \quad \underline{w}_C = \sum_{l=1}^{N_y-1} \beta_l \cdot \underline{\mu}_l.$$

Now the preconditioning operation $\underline{w}_C = C_C^{-1}\underline{d}_C$ can be rewritten as follows:

1. Express $\underline{d}_C$ in terms of eigenfrequencies μ_l and calculate the Fourier coefficients γ_l (*Fourier analysis*).

2. Divide these Fourier coefficients by the square root of the eigenvalues of K_y; i.e., $\beta_l := \gamma_l / \sqrt{\lambda_l(K_y)}$.

3. Calculate the preconditioned defect $\underline{w}_C$ by *Fourier synthesis*.

If we now denote the Fourier transformation by the square matrix $F = [\mu_l(i)]_{l,i=1}^{N_y-1}$, then the above procedure can be written as

$$C_C = 2\,K_y^{1/2} = \frac{2}{h} \begin{pmatrix} 2 & -1 & & & \\ -1 & 2 & -1 & & \\ & \ddots & \ddots & \ddots & \\ & & -1 & 2 & -1 \\ & & & -1 & 2 \end{pmatrix}^{1/2} = F^T \Lambda F,$$

where $\Lambda = \text{diag}\,[\lambda_l(C_C)]$, with $\lambda_l(C_C) = 2\sqrt{\lambda_l(K_y)}$. Due to the property $F = F^{-1} = F^T$ we can solve the system $C_C\underline{w}_C = F^T \Lambda F \underline{w}_C = \underline{d}_C$ in three simple steps:

$$\begin{aligned} \text{Fourier analysis:} \quad & \underline{\alpha} = F\underline{d}_C = F^T\underline{d}_C, \\ \text{scaling:} \quad & \underline{\beta} = \Lambda^{-1}\underline{\alpha}, \\ \text{Fourier synthesis:} \quad & \underline{w}_C = F\underline{\beta} = F^{-1}\underline{\beta}. \end{aligned}$$

Usually, the Fourier transformations in the Fourier analysis and the Fourier synthesis require a considerable amount of computing power, but this can be dramatically accelerated by the fast Fourier transformation (FFT), especially if N_y is a power of 2; i.e., $N_y = 2^r$. We refer to the original paper [23] and the book by W. Briggs and V. Henson [15] for more details.

2.4 Some other discretization methods

Throughout the remainder of this book we emphasize the finite element method (FEM). Before the FEM is defined in complete detail in Chapter 4, we want to show the reader one of the primary differences between it and the finite difference method (FDM). Both FDM and FEM offer radically different ways of evaluating approximate solutions to the original PDE at arbitrary points in the domain.

The FDM provides solution values at grid points, namely, $\{u_{i,j}\}$. Hence, the solution lies in a vector space, not a function space.

The FEM computes coefficients for basis functions and produces a solution function in a function space, not a vector space. The FEM solution u_h can be written as

$$u_h(x, y) = \sum u^{(i,j)} \varphi^{(i,j)}(x, y), \qquad (2.19)$$

where $\{u^{(i,j)}\}$ is computed by the FEM for a given set of basis functions $\{\varphi^{(i,j)}\}$.

To get the solution at a random point (x, y) in the domain using the FDM solution data $\{u_{i,j}\}$, there are two possibilities:

- If (x, y) lies on the grid (i.e., $x = x_i$ and $y = y_j$, for some i, j), then $u_h(x, y) = u_{ij}$.

- Otherwise, interpolation must be used. This leads to another error term, which may be larger than the truncation error of the FDM and lead to a much less accurate approximate solution than is hoped for.

For the FEM, the basis functions usually have very compact support. Hence, only a fraction of the $\{\varphi^{(i,j)}\}$ are nonzero. A bookkeeping procedure identifies nonzero $\varphi^{(i,j)}$ and evaluates (2.19) locally. For orthonormal basis sets, the evaluation can be particularly inexpensive. There is no interpolation error involved, so whatever the error is in the FEM is all that is in the evaluation anywhere in the domain.

We do not treat the finite volume methods (FVM) in this book. There are two classes of FVM:

- ones that are the box scheme composed with an FDM [62], and

- ones that are the box scheme composed with an FEM [6, 59].

In addition, we do not consider the boundary element method, which is also widely used in some applications [113].

Chapter 3

Introduction to Parallelism

The question of whether computers can think is just like the question of whether submarines can swim.
—Edsger W. Dijkstra (1930–2002)

3.1 Classifications of parallel computers

3.1.1 Classification by Flynn

In 1966, Michael Flynn [37] categorized parallel computer architectures according to how the data stream and instruction stream are organized, as is shown in Table 3.1.

Instruction Stream			
Single	Multiple		
SISD	**MISD**	Single	Data
SIMD	**MIMD**	Multiple	Stream

Table 3.1. *Classification by Flynn (Flynn's taxonomy).*

The multiple instruction, single data (MISD) class describes an empty set. The single instruction single data (SISD) class contains the normal single-processor computer with potential internal parallel features. The single instruction, multiple data (SIMD) class has parallelism at the instruction level. This class contains computers with vector units such as the Cray T-90, and systolic array computers such as Thinking Machines Corporation (TMC) CM2 and machines by MasPar. These parallel computers execute one program in equal steps and are not in the main scope of our investigations. TMC and MasPar expired in the mid-1990s. Cray was absorbed by SGI in the late 1990s. Like a phoenix, Cray rose from its ashes in 2001. Besides the recent processors by NEC and Fujitsu, even state-of-the-art PC processors by Intel and AMD contain a minor vector unit.

Definition 3.1 (MIMD). *MIMD means multiple instructions on multiple data and it characterizes parallelism at the level of program execution, where each process runs its own code.*

Usually, these MIMD codes do not run fully independently, so that we have to distinguish between *competitive processes* that have to use shared resources (center part of Fig. 3.1) and *communicating processes*, where each process possesses its data stream, which requires data exchange at certain points in the program (right part of Fig. 3.1).

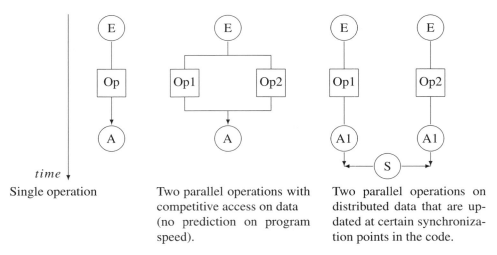

time

Single operation

Two parallel operations with competitive access on data (no prediction on program speed).

Two parallel operations on distributed data that are updated at certain synchronization points in the code.

Figure 3.1. *Types of parallel processes.*

Normally each process runs the same code on different data sets. This allows us to introduce a subclass of MIMD, namely, the class single program on multiple data (SPMD).

Definition 3.2 (SPMD). *The single program on multiple data programming model is the main class used in parallel programming today, especially if many processors are used and an inherent data parallelism is definable.*

There is no need for a synchronous execution of the codes at instruction level; thus an SPMD machine is *not* an SIMD machine. However, there are usually some synchronization points in a code to update data that are needed by other processors in order to preserve data consistency in implemented algorithms.

We distinguish between multiple processes and multiple processors. Processes are individual programs. There may be one or more processes running on one individual processor. In each process, there may be one or more threads. Threads are "light-weight" processes in the sense that they share all of the memory. While almost all of the algorithms described in this tutorial can be implemented using threads, we normally only assume parallelism at the processor level.

3.1.2 Classification by memory access

Definition 3.3 (Shared Memory). *Shared memory is memory that is accessed by several competitive processes "at the same time." Multiple memory requests are handled by hardware or software protocols.*

A shared memory parallel computer has the following advantage and disadvantages:

$\boxed{+}$ Each process has access to all of the data. Therefore, a sequential code can be easily ported to parallel computers with shared memory and usually leads to a first increase in performance with a small number of processors $(2, \ldots, 8)$.

$\boxed{-}$ As the number of processors increases, the number of memory bank conflicts and other access conflicts usually rises. Thus, the *scalability*, i.e., performance proportional to number of processors (i.e., wall clock time indirectly proportional to the number of processors) cannot be guaranteed unless there is a lot of memory and memory bandwidth.

$\boxed{-}$ Very efficient access administration and bus systems are necessary to decrease access conflicts. This is one reason these memory subsystems are more expensive.

We introduce briefly three different models to construct shared memory systems [64].

Definition 3.4 (UMA). *In the uniform memory access model for shared memory, all processors have equal access time to the whole memory, which is uniformly shared by all processors.*

UMA is what vendors of many small shared memory computers (e.g., IBM SP3, two- to four- processor Linux servers) try to implement. IBM and commodity Linux shared memory machines have not yet expired as of 2002.

Definition 3.5 (NUMA). *In the nonuniform memory access model for shared memory, the access time to the shared memory varies with the location of the processor.*

The NUMA model has become the standard model used in shared memory supercomputers today. Examples are those made today by HP, IBM, and SGI.

Definition 3.6 (COMA). *In the cache only memory access model for shared memory, all processors use only their local cache memory, so that this memory model is a special case of the NUMA model.*

The Kendall Square Research KSR-1 and KSR-2 had such a COMA memory model. Kendall Square expired in the mid-1990s.

Definition 3.7 (Distributed Memory). *Distributed memory is a collection of memory pieces where each of them can be accessed by only one processor. If one processor requires data* *stored in the memory of another processor, then communication between these processors (communicating processes) is necessary.*

We have to take into account the following items for programming on distributed memory computers:

$\boxed{+}$ There are no access conflicts between processors since data are locally stored.

$\boxed{+}$ The hardware is relatively inexpensive.

$\boxed{+}$ The code is potentially nearly optimally scalable.

$\boxed{-}$ There is no direct access to data stored on other processors and so communication via special channels (links) is necessary. Hence, a sequential code does not run and special parallel algorithms are required.

$\boxed{\cdot}$ The ratio between arithmetic work and communication is one criterion for the quality of a parallel algorithm. The time needed for communication is underestimated quite frequently.

$\boxed{\cdot}$ Bandwidth and transfer rate of the network between processors are of extreme importance and are always overrated.

Recent parallel computer systems by vendors such as IBM, SUN, SGI, HP, NEC, and Fujitsu are no longer machines with purely distributed or purely shared memory. They usually combine 2–16 processors into one computing node with shared memory for these processors. The memory of different computing nodes behaves again as distributed memory.

Definition 3.8 (DSM). *The distributed shared memory model is a compromise between shared and distributed memory models. The memory is distributed over all processors but the program can handle the memory as shared. Therefore, this model is also called the* virtual shared memory *model.*

In DSM, the distributed memory is combined with an operating system based on a message-passing system (see section 3.2.2) which simulates the presence of a global shared memory, e.g., a "sea of addresses" by the "interconnection fabric" of KSR and SGI.

$\boxed{+}$ The great advantage of DSM is that a sequential code runs immediately on this memory model. If the algorithms take advantage of the local properties of data, i.e., most data accesses of a process can be served from its own local memory, then good scalability can be achieved. On the other hand, the parallel computer can also be handled as a pure distributed memory computer.

For example, the SGI parallel machines Origin 2000 and Origin 3000 have symmetric multiprocessing (SMP). Each processor (or a small group of processors) has its own local memory. However, the parallel machine handles the whole memory as one huge shared memory. This realization was made possible by using a very fast crossbar switch (CrayLink). SGI was not expired as of 2002.

Besides hardware solutions to DSM, there are a number of software systems that simulate DSM by managing the entire memory using sophisticated database techniques.

Definition 3.9 (Terascale GRID Computers). *Several computer centers cooperate with huge systems, connected by very, very fast networks capable of moving a terabyte in a few seconds. Each node is a forest of possibly thousands of local DSMs. Making one of these work is a trial (comic relief?).*

3.1.3 Communication topologies

How can we connect the processors of a parallel computer? We are particularly interested in computers with distributed memory since clusters of fast PCs are really easy to construct and to use. A more detailed discussion of topologies can be found in [75].

Definition 3.10 (Link). *A link is a connection between two processors. We distinguish between a* unidirectional *link, which can be used only in one direction at a time, and a* bidirectional *link, which can be used in both directions at any time. Two unidirectional links are not a bidirectional link, however, even if the function is identical.*

Definition 3.11 (Topology). *An interconnection network of the processes is called a topology.*

Definition 3.12 (Physical Topology). *A physical topology is the interconnection network of the processors (nodes of the graph) given in hardware by the manufacturer.*

This network can be configured by changing the cable connections (such as IBM SP) or by a software reconfiguration of the hardware (e.g., Xplorer, MultiCluster-I). Recent parallel computers usually have a fixed connection of the processor nodes, but it is also possible to reconfigure the network for commonly used topologies by means of a crossbar switch that allows potentially all connections.

As an example, the now ancient Transputer T805 was a very specialized processor for parallel computing in the early 1990s. It had four hardware links so that it could handle a 2D (two-dimensional) grid topology (and a 4D hypercube) directly by hardware. Using one processor per grid point of the finite difference discretization of Poisson's equation in the square (2.4) in Chapter 2 resulted in a very convenient parallel implementation on this special parallel computer.

Definition 3.13 (Logical Topology). *The logical topology refers to how processes (or processor nodes) are connected to each other. This may be given by the user or the operating system. Typically it is derived from the data relations or communication structure of the program.*

The mapping from the logical to the physical topology is done via a parallel operating system or by parallel extensions to the operating system (e.g., see section 3.2.5). For example, a four-process program might use physical processors 1, 89, 126, and 1023 in a 1024-processor system. Logically, the parallel program would assume it is using processors

0 to 3. On most shared memory computers, the physical processors can change during the course of a long computation.

Definition 3.14 (Diameter). *The diameter of a topology is the maximal number of links that have to be used by a message sent from an arbitrary node p to another node q.*

The diameter is sometimes difficult to measure precisely. On some machines there is more than one path possible when sending a message between nodes p and q. The paths do not necessarily have the same number of links and are dependent on the network traffic load. In fact, on some machines, a message can fail by not being delivered in a set maximum period of time.

3.2   Specialties of parallel algorithms

3.2.1   Synchronization

Remark 3.15. Sequential programming is only the expression of our inability to transfer the natural parallelism of the world to a machine.

Administrating parallel processes provides several challenges that are not typical on single-processor systems. Synchronizing certain operations so that conflicts do not occur is a particularly hard problem that cannot be done with software alone.

Definition 3.16 (Undefined Status). *An undefined status occurs when the result of a data manipulation cannot be predicted in advance. It happens mostly when several processes have to access restricted resources, i.e., shared memory.*

process A: $N := N + 1$	clock.	process B: $N := N - 1$
LOAD N	(1)	LOAD N
INC N	(2)	DEC N
STORE N	(3)	STORE N

Figure 3.2. *Undefined status.*

For example, consider what happens when the two processes in Fig. 3.2 running simultaneously on different processors modify the same variable. The machine code instruction INC increments a register variable and DEC decrements it. The value of N depends on the execution speed of processes A and B and therefore its value is not predictable; i.e., the status of N is undefined, as shown in Fig. 3.3. In order to avoid programs that can suffer from data with an undefined status, we have to treat operations A and B as atomic operations; i.e., they cannot be split any further. This requires exclusive access to N for one process during the time needed to finish the operation. This exclusive access is handled by synchronization.

Definition 3.17 (Synchronization). *Synchronization prevents an undefined status of a*

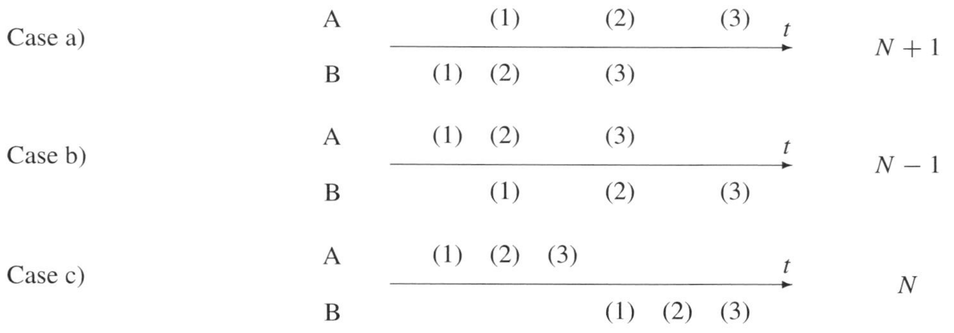

Case a)

A   (1)   (2)   (3) t

B   (1)   (2)   (3)

$N + 1$

Case b)

A   (1)   (2)   (3) t

B   (1)   (2)   (3)

$N - 1$

Case c)

A   (1)   (2)   (3) t

B   (1)   (2)   (3)

N

Figure 3.3. *Three possible outcomes.*

variable by means of a mechanism that allows processes to access that variable in a well-defined way.

Synchronization is handled by a semaphore mechanism (see [25, 41]) on shared memory machines and by message passing on distributed memory machines.

Definition 3.18 (Barrier). *The barrier is a special synchronization mechanism consisting of a certain point in a program that must be passed by all processes (or a group of processes) before the execution continues. This guarantees that each single process has to wait until all remaining processes have reached that program point.*

3.2.2 Message passing

Message passing is a mechanism for transferring data directly from one process to another. We distinguish between blocking and nonblocking communication (Figs. 3.4 and 3.5).

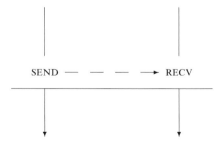

Figure 3.4. *Blocking communication.*

Blocking communication makes all (often two) involved processes wait until all processes signal their readiness for the data/message exchange. Usually the processes wait

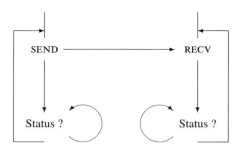

Figure 3.5. *Nonblocking communication.*

further until the exchange is completed.

Nonblocking communication allows all processes to send or receive their data/message totally independently of the remaining processes.

- This guarantees high efficiency in communication without deadlock (contrary to blocking communication, which can deadlock under very subtle conditions).

- The sending process expects no acknowledgment that the operation was executed successfully. Output parameters of the receiving process must not have an undefined value. Otherwise the whole program may crash. Preventing this deadlock is required from the programmer.

- Other operations can be executed between calling a communication routine and its appropriate, but not required, status query.

The pure semaphore concept is typical for shared memory computers and the resource management is usually unnoticed by the programmer. On the other hand, message passing occurs on distributed memory computers and must be specified explicitly by the programmer.

3.2.3   Deadlock

A general problem in the access to shared resources consists of preventing *deadlocks*.

Definition 3.19 (Deadlock). *A deadlock occurs when several processes are waiting for an event that can be released only by one of these waiting processes.*

The simplest example in which a deadlock may occur is the exchange of data between two processors if blocking communication is used. Here, both processes send their data and cannot proceed until both get an acknowledgment that the data have been received by the other process. But neither process can send that acknowledgment since neither process ever reaches that program statement. Therefore, each process has to wait for an acknowledgment from the other one and never reaches the RECV statement. One solution to overcome this deadlock is presented in section 3.4. Communication routines of more complexity than the

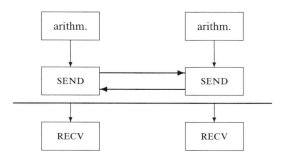

Figure 3.6. *Deadlock in blocking communication.*

simple send/receive between two processes in Fig. 3.6 might have deadlocks that are not as obvious.

3.2.4 Data coherency

Definition 3.20 (Data Coherency). *Data coherency is achieved if all copies of a data set always include the same data values.*

- On distributed memory machines, the programmer is responsible for data coherency.

- On classical (older) machines with huge shared memory, data coherency was guaranteed by semaphores. This limited the number of processors to about 40.

- Nowadays, shared memory computers have a local cache on each processor (sometimes a DSM model is used for programming). Here *cache coherency* is additionally required, which is much harder to implement than the simple coherency. Several mechanisms/protocols are used to guarantee *cache coherency*. In particular, the most recent DSM machines use an appropriate Cache Coherency NUMA model with internal protocols. Examples are as follows:

  snoopy-based protocol: Each process sends its memory accesses to the whole network. This leads to problems with the communication bandwidth.

  directory-based protocol: Each memory block (cache line) possesses an entry in a table, which stores the state of that block, and a bit vector which includes all processes with copies of that block. It can be determined by analysis of these entries which caches have to be updated [96].

3.2.5 Parallel extensions of operating systems and programming languages

In the early years of parallel programming, mostly in research, only a few hardware-dependent calls were available. As a consequence, source code was not portable and one had to create a hardware-independent interface of one's own. Nowadays, all vendors of parallel computers offer standard parallel libraries and/or similar extensions of the operating system, which include all necessary compilers, libraries, and so on.

Some parallel versions of high-level programming languages are available, e.g., High Performance Fortran (HPF[1]), Vienna Fortran,[2] and OpenMP[3] [21] extensions to C, C++, and Fortran. They all include language extensions that support parallelism on structured data and their compilers try an automatic parallelization.

Another, and more promising, approach consists of the development of general and portable interfaces for multiple instructions on multiple data (MIMD) computers. The most common library in this field is the message-passing interface (MPI), which is based on earlier noncommercial and commercial interfaces like PVM, Express, etc. The MPI[4] consortium is supported by all major manufacturers of parallel machines (they work actively in that committee) and the library is available on all software platforms, e.g., LINUX,[5] and even on some variants of Windows (e.g., 2000 and XP). Therefore, we focus in the course material on MPI. Section 3.3 is strongly recommended for a deeper understanding of the calls in MPI;[6] see also [38, 49, 84].

3.3   Basic global operations

3.3.1   SEND and RECV

Each parallel program needs basic routines for sending and receiving data from one process to another. We use symbolic names for these routines that can be mapped to concrete implementations. Let us introduce the routine

$$\text{SEND}(nwords, data, ProcNo)$$

for sending $nwords$ of data stored in $data$ from the calling process to process $ProcNo$. The appropriate receiving routine is

$$\text{RECV}(nwords, data, ProcNo).$$

The specification of what sort of data (DOUBLE, INTEGER, etc.) is transferred is of no interest for our general investigations. These details have to be taken into account by choosing the right parameters in the point-to-point communication of a given parallel library. We use exclusively the blocking communication, implemented as MPI_SEND and MPI_RECV in the MPI library. Using only the blocking versions of the communication routines assists a beginner in parallel programming in such a way that only one message thread has to be debugged at a certain program stage. Once the code runs perfectly it can be accelerated by using nonblocking communication.

---

[1] http://www.mhpcc.edu/training/workshop/hpf/MAIN.html
[2] http://www.vcpc.univie.ac.at/activities
[3] http://www.openmp.org
[4] http://www.mcs.anl.gov/mpi/index.html
[5] http://www.suse.com
[6] http://nexus.cs.usfca.edu/mpi

3.3.2   EXCHANGE

The routines SEND and RECV can be combined into a routine

$$\text{EXCHANGE}(ProcNo, nWords, SendData, RecvData),$$

which exchanges data between processes p and q.

Nonblocking communication

The variant in Fig. 3.7 works *only* in the case of nonblocking communication; i.e., the sending process does not require an acknowledgment that the message was received. Use of blocking communication ends in deadlock.

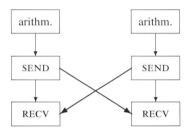

Figure 3.7. *Nonblocking* EXCHANGE.

Blocking communication

In the case of a blocking communication, a deadlock-free EXCHANGE between p and q needs a unique function determining which one of the two processes may send/receive first. The function $\text{TEST}(p, q) = (p > q)$ provides an ordering mechanism such that both processes achieve the same result independently. The definition that the process with the larger process number first sends and then receives (vice versa on the other process) guarantees a deadlock-free EXCHANGE procedure (Fig. 3.8). An appropriate MPI call is MPI_SENDRECV, which determines p and q automatically.

3.3.3   Gather-scatter operations

A parallel program usually possesses a *root* process, which has exclusive access to resources. This process has to gather and scatter specific data from and to the other processes. In the function

$$\text{SCATTER}(root, SendData, MyData),$$

each process (including root) receives from the root process its specific data *MyData* out of the global data set *SendData* stored only at the root process. On the other hand, the function

$$\text{GATHER}(root, RecvData, MyData)$$

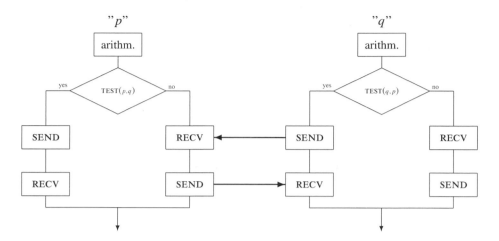

Figure 3.8. *Blocking* EXCHANGE.

collects the specific data in *MyData* on the root process and stores it in *RecvData*. Both functions can be implemented via SEND or RECV in various algorithms. Appropriate MPI calls are MPI_GATHER and MPI_SCATTER (see Fig. 3.9).

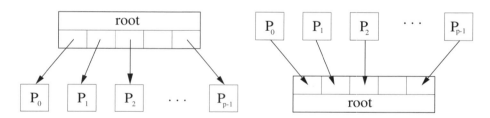

Figure 3.9. SCATTER *and* GATHER.

Sometimes it is necessary to gather and scatter data from/to all processes. The simplest realization of the routines GATHER_ALL and SCATTER_ALL can be implemented by using combined GATHER/SCATTER calls.

3.3.4   Broadcast

Often, all processors receive the same information from one process (or all processes). In this case the *root* process acts like a broadcast station and the appropriate routine is

$$\text{BROADCAST}(root, nWords, Data).$$

This is a simplification of the SCATTER routine, which distributes individual data to the processes. The MPI call is MPI_BCAST (see Fig. 3.10).

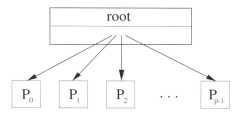

Figure 3.10. BROADCAST *operation.*

3.3.5 Reduce and reduce-all operations

Here we perform an arithmetical/logical operation on the distributed data of all processes (e.g., $s = \sum_{i=0}^{p-1} a_i$). At the end, the result is available on one root process (REDUCE operation) or on all processes (REDUCE_ALL). The function could look like REDUCE(n, X, Y, $VtOp$) and it can handle vectors and operations $VtOp$ on them in the parameter list. So the above sum s could be calculated via REDUCE(`1,S,A,Double_Plus`).

The simplest implementation of REDUCE consists of a GATHER followed by DOU-BLE_PLUS on the root process and then a SCATTER. More sophisticated and faster implementations can be derived for certain topologies. The calls MPI_REDUCE and MPI_ALLREDUCE are the appropriate functions in the MPI library.

3.3.6 Synchronization by barriers

The synchronization of all processes is called a barrier and is provided by MPI as MPI_BARRIER.

- Each operation in sections 3.3.2 to 3.3.5 can be used as a barrier; e.g., we can construct a barrier via REDUCE(1, I, J, Int_Plus).

- Most parallel algorithms quite naturally contain a barrier, since a global data exchange (e.g., summation) is necessary at some point in the algorithm.

3.3.7 Some remarks on portability

Operations in sections 3.3.2 to 3.3.6 have been built on the following conditions.

1. There is a given topology, e.g., farm, binary tree, or hypercube [75, 89]. If the assumed or required topology is not physically available, then it must be constructed logically on the given topology.

2. Send to a process: SEND.

3. Receive from a process: RECV.

Only these three items depend on the hardware and operating system. All remaining routines should be based on these three basic routines.

Parallel communication libraries come and go. TCP/IP, UDP, PVM, P4, TCMSG, and MPI are just a few of the once popular libraries. Before writing any parallel application, either get a portable, wrapper library from a friend or colleague or write your own. Do not hard-code any particular library calls into your code. This simplifies the code for debugging and make it more portable. Thus, a parallel program is trivially slower than a hardware-optimized parallel code in real production runs of very large applications.

3.4   Performance evaluation of parallel algorithms

How do we compare fairly the suitability or lack thereof of parallel algorithms for large or huge numbers of processors?

3.4.1   Speedup and scaleup

Definition 3.21 (Scalability). *The property of a program to adapt automatically to a given number of processors is called scalability.*

Scalability is more sought after than highest efficiency (i.e., gain of computing time by parallelism) on any specific architecture/topology.

Definition 3.22 (Granularity). *Granularity is a measure of the size of program sections that are executable without communication with other processes. The two extremes are* fine grain *and* coarse *grain algorithms.*

Definition 3.23 (Speedup). *The speedup S_P is a measure of the performance gain of a parallel code running on P processors in comparison with the sequential program version. The global problem size N_{glob} remains unchanged. This measure can be expressed quantitatively by*

$$\boxed{S_P \;=\; \frac{t_1}{t_P}},\tag{3.1}$$

where t_1 is the system time of process A on one processor and t_P is the system time of process A on P processors.

A speedup of P is desirable; i.e., P processes are P times as fast as one. The global problem size N_{glob} is constant and therefore the local problem size on each processor decreases as $N_i \sim \frac{N_{glob}}{P} \ \forall\, i = \overline{1,P}$; i.e., the more processors that are used, the less an individual processor has to do.

S_P can be calculated in a variety of ways. Our speedup measures only the parallelization quality of *one* algorithm that is slightly adapted to P. The quality of the numerical algorithm itself is measured with the numerical efficiency in (3.6). Only the combination of the two measures allows a fair evaluation of an algorithm.

Example 3.24 (Speedup Worker). *Let us use workers instead of processors. $P = 1$ worker digs a hole of $1\,m^3$ [N_{glob}] in $t_1 = 1$ h by stepping into the pit, taking a shovelful of soil, and dropping it a few meters away. Two workers can reduce the time for digging that hole* *in the described way by a factor of 2; i.e., it takes them* $t_2 = 0.5$ *h. How long [t_P] does it take* $P = 1000$ *workers to perform the same work?*

Obviously, the workers obstruct each other and too high of an organization overhead develops. Therefore, we gain very little speedup when using many workers.

This example leads us directly to Amdahl's theorem.

Theorem 3.25 (Amdahl's Theorem (1967)). *Each algorithm contains parts that cannot be parallelized.*
Let s be the sequential part of an algorithm and p the parallel part of an algorithm (both normalized):

$$s + p = 1. \tag{3.2}$$

Thus, we can express the system time on one processor by $t_1 \longrightarrow s + p$ *and (under the assumption of an optimal parallelization) the system time on P processors by* $t_P \longrightarrow s + \frac{p}{P}$. *This results in an upper barrier for the maximally attainable speedup:*

$$\boxed{S_{P,max} = \frac{s + p}{s + \frac{p}{P}} = \frac{1}{s + \frac{1-s}{P}} \leq \frac{1}{s}.} \tag{3.3}$$

If we can parallelize 99% of a code, then the remaining 1% sequential code restricts the maximal parallel performance such that a maximal speedup $S_{\infty,max} = 100$ can be achieved by formula (3.3).

P	10	100	1000	10000
S_P	9	50	91	99

The use of more than 100 processors appears senseless, since the cost of 1000 processors is (approximately) 10 times as high as the cost of 100 processors but the performance does not even double. This result seems to be discouraging.

But, is Amdahl's theorem suitable for us?

There is an observation that certain algorithms achieve a speedup even higher than P on smaller processor numbers (<20). This is usually caused by a more intensive use of cache memory on the processors if the size of the local problem is getting smaller or by a different number of iterations per processor in a domain decomposition (DD) algorithm.

On the other hand, Amdahl's theorem does not take into account any time for communication, so that the real behavior of the speedup is always worse; see Fig. 3.11 for a qualitative illustration. We can formulate the consequence that the attainable speedup depends on the size and communication properties of the underlying global problem.

Example 3.26 (Scaleup Worker). *If each worker digs a pit of 1 m^3, then 1000 workers reach nearly 1000 times as high performance during that hour. They all work on one huge project.*

Usually the sequential part of a code is rather insensitive to the overall problem size. On the other hand, the parallel parts cover most of the arithmetic in a code. Therefore, increasing the time a code spends in the parallel parts by increasing the problem size decreases

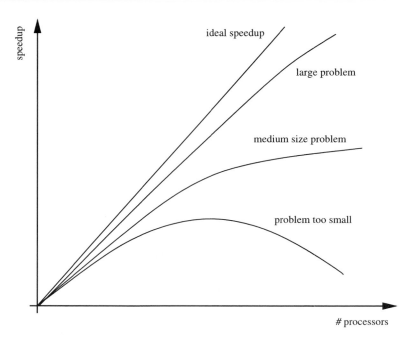

Figure 3.11. *Speedup with respect to problem size.*

the importance of the sequential part. In conclusion, the scaling of the problem size with
the number of processors is a way to overcome Amdahl's theorem.

Definition 3.27 (Scaled Speedup). *The scaled speedup S_C is a measure of the parallel per-formance gain when the global problem size N_{glob} increases with the number of processors used. It assumes additionally that the parallel part of the algorithm is optimal with respect to the problem size, i.e., $O(N_{glob})$. We can express the scaled speedup quantitatively:*

$$S_C(P) \;=\; \frac{s_1 + P \cdot p_1}{s_1 + p_1} \overset{s_1+p_1=1}{=} s_1 + P(1 - s_1),\tag{3.4}$$

by using the notation $s_1 + p_1$ for the normalized system time on a parallel computer and $s_1 + P \cdot p_1$ for the system time on a sequential computer.

Equation (3.4) means that a sequential part of 1% leads to a scaled speedup of $S_C(P) \approx P$ because the serial part decreases with the problem size. This theoretical forecast is confirmed in practice.

The (scaled) speedup may not be the only criterion for the usefulness of an algorithm;
e.g., one can see in Fig. 3.12 (courtesy of Matthias Pester, Chemnitz, 1992), by comparing
Gaussian elimination with a diagonally preconditioned conjugate gradient (PCG), that

$$S_P(\text{Gauss}) > S_P \,(\text{PCG}) \quad BUT \quad t(\text{Gaussian}) >> t(\text{PCG});$$

i.e., a good parallelization is worthless for a numerically inefficient algorithm.

If one compares algorithms implemented on different parallel machines, then the costs of that machine should also be taken into account, e.g., price per unknown variable.

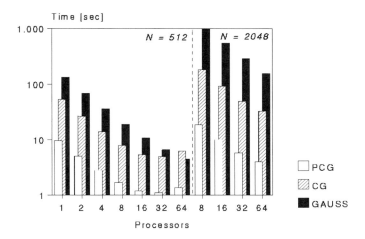

Figure 3.12. *System times for Gaussian elimination, CG, PCG.*

Remark 3.28. Some people only want to measure flops (floating point operations per second), others only want to measure speedup. Actually, only wall clock time counts for a parallel code. If you do not want the solution *now*, why bother with the nuisance of parallel programming?

3.4.2 Efficiency

Definition 3.29 (Parallel Efficiency). *The parallel efficiency tells us how close our parallel implementation is to the optimal speedup. Here we present only the formula for the scaled parallel efficiency. The classical efficiency can be calculated similarly.*

$$E_{C,par} = \frac{S_C(P)}{P}. \tag{3.5}$$

A desirable efficiency would be 1 (= 100%); i.e., the use of P processors accelerates the code by a factor of P. The parallel efficiency for the scaleup can be given explicitly:

$$E_{C,par} = \frac{S_C(P)}{P} = \frac{s_1 + P(1 - s_1)}{P} = 1 - s_1 + \frac{s_1}{P} > 1 - s_1.$$

We get a slightly decreasing efficiency with increasing number of processors but the efficiency is, however, at least as high as in the parallel part of the program.

Definition 3.30 (Numerical Efficiency, Scaled Efficiency).   *The numerical efficiency compares the fastest sequential algorithm with the fastest parallel algorithm implemented on one processor via the relationship*

$$E_{num} = \frac{t_{parallel}}{t_{serial}}. \tag{3.6}$$

From this follows the scaled efficiency of a parallel algorithm:

$$E = E_{C,par} \cdot E_{num}. \tag{3.7}$$

Formula (3.7) does not contain any losses due to communication and it also assumes a uniform distribution of the global problem on all processors. This uniform distribution cannot be achieved in practice and the workload per processor can change very often during nonlinear and/or adaptive calculations. Therefore, the attained efficiency is lower than the theoretical one, often vastly so. One way to obtain a good theoretical efficiency in practice consists of the use of load-balancing strategies.

Definition 3.31 (Load Balance). *The equal distribution of the computational workload on all processors is called load balance.*

Opportunities to achieve a load balance close to the optimum follow:

- Static load balancing can be done a priori:

    - Simple distribution of meshes or data based on an arbitrary numbering is not efficient for the parallel code since it does not take advantage of spatial data locality.

    - Distribution of data with several bisection techniques (e.g., coordinate bisection, recursive spectral bisection [7], Kerninghan–Lin algorithm [102]) produces much better data splittings with respect to communication. The better the desired data distribution, the more expensive the individual bisection techniques become.

      Static load balancing is especially suited to optimization problems or the distribution of coarse meshes/grids in multigrid algorithms.

- Dynamic load balancing contains a redistribution of data during the calculations [9, 111]:

    - A great deal of effort for the load distribution is needed and good heuristics are necessary that avoid a communication overhead.

    - Dynamic load balancing is required in highly adaptive algorithms.

Remark 3.32. METIS, PARMETIS, JOSTLE, and PARTI are the most commonly used load-balancing codes. Many others exist, however.

3.4.3 Communication expenditure

We mostly neglected the time needed for communication (exchange of data) in all investigations of speedup and efficiency.

Let t_{Startup} be the latency (time until communication starts) and t_{Word} the time for transferring one Word (determined by bandwidth). Then the time to transfer N Words together (message) is

$$t_N = t_{\text{Startup}} + N \cdot t_{\text{Word}}. \tag{3.8}$$

Regarding (3.8) and taking into account practical experience, the transfer of *one* long message is faster than the transfer of several short messages. The operating system and/or hardware splits huge messages into several smaller ones. The size of these smaller messages is vendor/OS dependent, but is usually > 1 kB. Table 3.2 represents a collection of latency time and bandwidth performance. The sustained bisection bandwidth is the bandwidth that can be achieved when one half of the processors communicates with the other half. As an example, the bisection bandwidth of the Origin2000 scales linearly up to 10 GB/s on 32 processors. This value remains the same for 64 processors and doubles to 20 GB/s with 128 processors.

Machine/OS	Latency	Bandwidth	
Xplorer/Parix	100 μs	1.3 MB/s	(measured)
Intel iPsc/860	82.5 μs	2.5 MB/s	
Convex SPP-1	0.2 μs	16.0 MB/s	
nCube2		71 MB/s	
Cray TT3D	6 μs	120 MB/s	
Intel Paragon		200 MB/s	
SGI Origin2000 R10k/250		680 MB/s	
HP 9000 N-class	105 ns	160 MB/s	(in Hyperfabric)
HP Superdome	360 ns	12.8 GB/s	
IBM RS/6000	1000 ns	0.5 GB/s	
SGI Origin3800	45 ns	1.6 GB/s	
SUN E10000	500 ns	12.8 GB/s	

Table 3.2. *Comparison of latency and bandwidth.*

Exercises

The following MPI functions require a *communicator* as parameter. This communicator describes the *group* of processes that are addressed by the corresponding MPI function. By default, all processes are collected in *MPI_COMM_WORLD*, which is one of the constants supplied by MPI. We restrict the examples to these global operations. For this purpose, create the special MPI-type variable

MPI_Comm icomm = MPI_COMM_WORLD,

which is used as parameter.

E3.1. Compile the program in *Example/firstc* via `make`. Start the program.

E3.2. Write your first parallel program by using

MPI_Init and **MPI_Finalize**,

compile the program with `make`, and start four processes via `mpirun -np 4`.

E3.3. Determine the number of your parallel processes and the local process rank by using
the routines

MPI_Comm_rank and **MPI_Comm_size**.

Let the root process (rank = 0) write the number of running processes. Start with
different numbers of processes.

E3.4. The file *greetings.c* includes a routine

Greetings(myid, numprocs, icomm)

that prints the names of the hosts your processes are running on. Call that routine
from your main program and change the routine such that the output is ordered with
respect to the process ranks. Study the routines

MPI_Send and **MPI_Recv**.

E3.5. Write a routine

Send_ProcD(to, nin, xin, icomm)

that sends *nin* double precision numbers of the array *xin* to the process *to*. Note that
the receiving process *to* does not normally have any information about the length of
the data it receives.

   Also write a corresponding routine

Recv_ProcD(from, nout, xout, maxbuf, icomm)

that receives *nout* double precision numbers of the array *xout* from the process *from*.
A priori, the receiving process does not have any information about the length of the
data to be received; i.e., *nout* is an output parameter. *maxbuf* stands for the maximum
length of the array *xout*.

E3.6. First test the routines from **E3.5** with two processes. Let process 1 send data and
process 0 receive them. Extend the test to more processes.

E3.7. Combine the routines from **E3.5** into one routine

ExchangeD(yourid, nin, xin, nout, xout, maxbuf, icomm),

which exchanges double precision data between your own process and another process *yourid*. The remaining parameters are the same as in **E3.5**. Test your routines
first with two and then with more processes.

Chapter 4

Galerkin Finite Element Discretization of Elliptic Partial Differential Equations

If people do not believe that mathematics is simple, it is only because they do not realize how complicated life is.
—John von Neumann (1903–1957)

In Chapter 2, we considered the Poisson equation and the finite difference method (FDM) as the simplest method for its discretization. In this chapter, we give a brief introduction to the finite element method (FEM), which is the standard discretization technique for elliptic boundary value problems (BVPs). As we will see, the FEM is nothing more than a Galerkin method with special basis functions. In contrast to the FDM, where we used the classical formulation of the elliptic BVP as a starting point for the discretization, the Galerkin FEM starts with the variational, or weak, formulation of the BVP that we want to solve.

4.1 Variational formulation of elliptic BVPs

In this section, we derive the variational (weak) formulation of some model BVPs for a scalar elliptic second-order partial differential equation (PDE). The same procedure is used for elliptic PDE systems (e.g., the linear elasticity problem) and for higher-order equations (e.g., the biharmonic problem).

The abstract form of the variational formulation of any linear elliptic problem in divergence form reads as follows:

$$\text{Find } u \in V_g: \quad a(u, v) = \langle F, v \rangle \quad \forall v \in V_0, \tag{4.1}$$

with a given continuous (= bounded) bilinear form $a(\cdot, \cdot) : V \times V \to \mathbb{R}^1$ defined on some Hilbert space V equipped with the scalar product $(\cdot, \cdot)$ and the corresponding norm $\| \cdot \|$, and a given continuous linear functional $\langle F, \cdot \rangle : V_0 \to \mathbb{R}^1$ defined on some subspace V_0 of V. We search for the solution u in the set $V_g := g + V_0 \equiv \{v \in V : \exists \, w \in V_0, \, v = g + w\}$ of admissible functions, which is usually a linear manifold defined by the inhomogeneous ($g \neq 0$) Dirichlet boundary conditions (BCs). As we will see, integration by parts is the basic technique for deriving the variational formulation from the classical statement of the BVP.

Let us first consider the *classical formulation* of a mixed BVP for a second-order linear elliptic PDE in the divergence form.

Find $u \in X := C^2(\Omega) \cap C^1(\Omega \cup \Gamma_2 \cup \Gamma_3) \cap C(\Omega \cup \Gamma_1)$ such that the PDE

$$-\sum_{i,j=1}^{m} \frac{\partial}{\partial x_i}\left(a_{ij}(x)\,\frac{\partial u}{\partial x_j}\right) + \sum_{i=1}^{m} a_i(x)\,\frac{\partial u}{\partial x_i} + a(x)u(x) = f(x) \qquad (4.2)$$

holds for all $x \in \Omega$ and that the BCs

- $u(x) = g_1(x) \ \forall \, x \in \Gamma_1$ (Dirichlet (1st-kind) BC),
- $\frac{\partial u}{\partial N} := \sum_{i,j=1}^{m} a_{ij}(x)\frac{\partial u(x)}{\partial x_j}\, n_i(x) = g_2(x) \ \forall \, x \in \Gamma_2$
  (Neumann (2nd-kind) BC),
- $\frac{\partial u}{\partial N} + \alpha(x)u(x) = g_3(x) \ \forall \, x \in \Gamma_3$ (Robin (3rd-kind) BC)

are satisfied.

The computational domain $\Omega \subset \mathbb{R}^m$ ($m = 1, 2, 3$) is supposed to be bounded and the boundary $\Gamma = \partial\Omega = \bar{\Gamma}_1 \cup \bar{\Gamma}_2 \cup \bar{\Gamma}_3$ should be sufficiently smooth. Throughout this chapter, $n = (n_1, \ldots, n_m)^T$ denotes the outer unit normal. $\frac{\partial u}{\partial N}$ is sometimes called the conormal derivative. It is clear that the Dirichlet problem (2.1) for the Poisson equation discussed in Chapter 2 is a special case of (4.2). The stationary heat conduction equation is another special case of (4.2); see Example 4.3.

Remark 4.1.

1. $u \in X$ solving (4.2) is called the classical solution of the BVP (4.2).

2. Note that the data $\{a_{ij}, a_i, a, \alpha, f, g, \Omega\}$ should satisfy classical smoothness assumptions such as $a_{ij} \in C^1(\Omega) \cap C(\Omega \cup \Gamma_2 \cup \Gamma_3)$, etc.

3. We assume uniform ellipticity of (4.2) in Ω; i.e.,

   - $a_{ij}(x) = a_{ji}(x) \quad \forall \, x \in \overline{\Omega}, \quad \forall \, i, j = \overline{1, m}$,
   - $\exists \, \bar{\mu}_1 = const. > 0$ such that
     $\Lambda(x, \xi) := \sum_{i,j=1}^{m} a_{ij}(x)\xi_i\,\xi_j \geq \bar{\mu}_1 |\xi|^2 \quad \forall \, \xi \in \mathbb{R}^m, \quad \forall \, x \in \overline{\Omega}$.

4. The Robin BC is often written in the form

$$-\frac{\partial u}{\partial N} = \alpha(x)(u(x) - u_e(x)),$$

   where $g_3(x) = \alpha(x)u_e(x)$, with given smooth functions α and u_e.

As mentioned in Remark 4.1, we have to assume that the data of the BVP (4.2) are smooth. However, in practice, the computational domain Ω is usually composed of several subdomains with different material properties. In this case, the differential equation is only valid in the subdomains where the data are supposed to be smooth (e.g., constant). On the interfaces between these subdomains, we have to impose interface conditions forcing the function u (the temperature in the case of a heat conduction problem) and its conormal derivative (the negative heat flux) to be continuous. The variational formulation that we are going to study now will automatically cover this practically important case. A heat conduction problem of this type will be introduced shortly and will be used as a model problem throughout this chapter.

So let us first derive the *variational formulation* of the BVP (4.2). The formal procedure for the derivation of the variational formulation consists of the following five steps:

1. Choose the space of test functions $V_0 = \{v \in V = H^1(\Omega) : v = 0 \text{ on } \Gamma_1\}$, where $V = H^1(\Omega)$ is the basic space for scalar second-order PDEs. The Sobolev space $H^1(\Omega)$, sometimes also denoted by $W_2^1(\Omega)$, consists of all quadratically integrable functions with quadratically integrable derivatives and is equipped with the scalar product

$$(u, v)_{1,\Omega} := \int_\Omega (uv + \nabla^T u \nabla v) dx \quad \forall\, u, v \in H^1(\Omega)$$

and the corresponding norm $\| \cdot \|_{1,\Omega} := \sqrt{(\cdot, \cdot)_{1,\Omega}}$ (see [1, 22] for more details about Sobolev spaces).

2. Multiply the PDE (4.2) by test functions $v \in V_0$ and integrate over Ω:

$$\int_\Omega \left(-\sum_{i,j=1}^m \frac{\partial}{\partial x_i} \left(a_{ij} \frac{\partial u}{\partial x_j} \right) + \sum_{i=1}^m a_i \frac{\partial u}{\partial x_i} + au \right) v \, dx = \int_\Omega fv \, dx \quad \forall\, v \in V_0.$$

3. Use integration by parts:

$$\int_\Omega \frac{\partial w}{\partial x_i} v \, dx = -\int_\Gamma w \frac{\partial u}{\partial x_i} dx + \int_{\partial \Omega} wn \cdot n_i \, ds$$

in the principal term (= term with the second-order derivatives):

$$\int_\Omega \left(\sum_{i,j=1}^m a_{ij} \frac{\partial u}{\partial x_j} \frac{\partial v}{\partial x_i} + \sum_{i=1}^m a_i \frac{\partial u}{\partial x_i} v + auv \right) dx - \int_{\partial \Omega} \underbrace{\sum_{i,j=1}^m a_{ij} \frac{\partial u}{\partial x_j} n_i}_{} \cdot v \, ds$$

$$=: \frac{\partial u}{\partial N} = \text{conormal derivative}$$

$$= \int_\Omega fv \, dx \quad \forall\, v \in V_0.$$

4. Incorporate the natural (Neumann and Robin) BCs on Γ_2 and Γ_3:

$$\int_\Gamma \frac{\partial u}{\partial N} v \, ds = \int_{\Gamma_1} \frac{\partial u}{\partial N} v \, ds + \int_{\Gamma_2} g_2 v \, ds + \int_{\Gamma_3} (g_3 - \alpha u) v \, ds.$$

5. Define the linear manifold, where the solution u is searched for:

$$V_g = \{v \in V = H^1(\Omega) : v = g_1 \text{ on } \Gamma_1\}.$$

Summarizing, we arrive at the following variational formulation.

> Find $u \in V_g$ such that $a(u, v) = \langle F, v \rangle \quad \forall v \in V_0$, where
>
> $$a(u, v) := \int_\Omega \left(\sum_{i,j=1}^m a_{ij} \frac{\partial u}{\partial x_j} \frac{\partial v}{\partial x_i} + \sum_{i=1}^m a_i \frac{\partial u}{\partial x_i} v + auv \right) dx + \int_{\Gamma_3} \alpha u v \, ds,$$
>
> $$\langle F, v \rangle := \int_\Omega fv \, dx + \int_{\Gamma_2} g_2 v \, ds + \int_{\Gamma_3} g_3 v \, ds,$$
>
> $$V_g := \{v \in V = H^1(\Omega) : v = g_1 \text{ on } \Gamma_1\},$$
>
> $$V_0 := \{v \in V : v = 0 \text{ on } \Gamma_1\}.$$
>
> (4.3)

Remark 4.2.

1. The solution $u \in V_g$ of (4.3) is called the weak or generalized solution.

2. For the variational formulation (4.3), the assumptions imposed on the data can be weakened (the integrals involved in (4.3) should exist):

   i) $a_{ij}, a_i, a \in L_\infty(\Omega)$, $\alpha \in L_\infty(\Gamma_3)$, i.e., bounded,

   ii) $f \in L_2(\Omega)$, $g_i \in L_2(\Gamma_i)$, $i = 2, 3$,

   iii) $g_1 \in H^{\frac{1}{2}}(\Gamma_1)$; i.e., $\exists \tilde{g}_1 \in H^1(\Omega) : \tilde{g}_1|_{\Gamma_1} = g_1$,

   iv) $\Omega \subset \mathbb{R}^m$ is supposed to be bounded with a Lipschitz-continuous boundary $\Gamma = \partial\Omega \in C^{0,1}$,   (4.4)

   v)   uniform ellipticity; i.e., $\exists \bar{\mu}_1 = const. > 0$ such that

   $$\sum_{i,j=1}^m a_{ij}(x)\xi_i \, \xi_j \geq \bar{\mu}_1 |\xi|^2 \quad \forall \xi \in \mathbb{R}^m,$$
   $$a_{ij}(x) = a_{ji}(x) \quad \forall i, j = \overline{1, m},$$
   a.e. $\in \Omega$.

Example 4.3 (The Plain Heat Conduction Problem CHIP). *Find the temperature field* $u \in V_g$ *such that the variational equation*

$$a(u, v) = \langle F, v \rangle \quad \forall v \in V_0$$

holds, where

$$a(u, v) := \int_\Omega \left(\lambda \frac{\partial u(x)}{\partial x_1} \frac{\partial v(x)}{\partial x_1} + \lambda \frac{\partial u(x)}{\partial x_2} \frac{\partial v(x)}{\partial x_2} + a(x)u(x)v(x) \right) dx + \int_{\Gamma_3} \alpha u v \, ds,$$

$$\langle F, v \rangle := \int_\Omega fv \, dx + \int_{\Gamma_2} g_2 v \, ds + \int_{\Gamma_3} g_3 v \, ds,$$

$$V_g := \{ v \in V = H^1(\Omega) : v = g_1 \; on \; \Gamma_1 \},$$

$$V_0 := \{ v \in V = H^1(\Omega) : v = 0 \; on \; \Gamma_1 \}.$$

The computational domain Ω consists of the two subdomains Ω_I and Ω_{II} with different material properties (see Fig. 4.1). In the silicon subdomain Ω_I, $\lambda = \lambda_I = 1 \; W(mK)^{-1}$ (heat conduction coefficient), $a = 0$ (no heat transfer through the top and bottom surfaces), and $f = 0$, whereas, in the copper subdomain Ω_{II}, $\lambda = \lambda_{II} = 371 \; W(mK)^{-1}$, $a = 0$, and $f = 0$; i.e., the domain data are piecewise constant. The boundary data on Γ_1, Γ_2, and Γ_3 (see Fig. 4.1 for the precise location of the boundary pieces) are prescribed as follows:
  $g_1 = 500 \; K$ *(prescribed temperature on Γ_1)*,
  $g_2 = 0$, *i.e., heat flux isolation on Γ_2*,
  $g_3 = \alpha u_e$ *(cf. Remark 4.1)*,
  $\alpha = 5.6 \; W(m^2 K)^{-1}$ *(heat transfer coefficient)*,
  $u_e = 300 \; K$ *(exterior temperature)*.

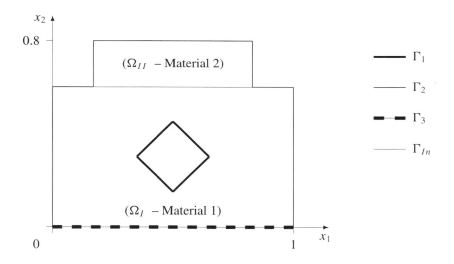

Figure 4.1. *Computational domain Ω consisting of two different materials.*

It is clear that the variational formulation of the problem CHIP given in Example 4.3 fulfills the requirements formulated in Remark 4.2. The solution $u \in V_g$ of the variational problem (4.1) has the form $u = g + w$, with some given admissible function $g \in V_g$ (e.g., $g = 500$ and some unknown function $w \in V_0$). Therefore, we can rewrite (4.1) in the following form:

$$\text{Find } w \in V_0 : \quad a(w, v) = \langle F, v \rangle - a(g, v) \quad \forall v \in V_0. \tag{4.5}$$

This procedure is called homogenization of the essential BCs. Indeed, for our variational problem (4.3), it is sufficient to find some function $g \in H^1(\Omega)$ satisfying the Dirichlet BCs

$g = g_1$ on Γ_1. This is always possible if the function g_1 is from $H^{1/2}(\Gamma_1)$ (see, e.g., [1, 22]). Then the function w is sought in V_0 such that (4.5) is satisfied, where the bilinear form and the linear form are defined as in (4.3). The homogenized variational formulation (4.5) simplifies the existence and uniqueness analysis. For simplicity, let us denote the right-hand side of (4.5) again by $\langle F, v \rangle$. Therefore, for the time being, we consider the variational problem in homogenized form:

$$\text{Find } w \in V_0 : \quad a(w, v) = \langle F, v \rangle \quad \forall\, v \in V_0, \tag{4.6}$$

where $a(\cdot, \cdot) : V \times V \to \mathbb{R}^1$ and $F \in V_0^*$ are supposed to be a given continuous (= bounded) bilinear form and a given linear continuous (= bounded) functional, respectively. Let us now consider the homogenized variational problem (4.6) under the following *standard assumptions*:

$$
\left.
\begin{array}{ll}
1. & F \in V_0^* \text{ (space of linear, continuous functionals)}, \\
2. & a(\cdot, \cdot) : V_0 \times V_0 \to \mathbb{R}^1 \text{—bilinear form on } V_0, \\
2.\,\text{a)} & V_0\text{—elliptic} : \mu_1 \|v\|^2 \le a(v, v) \quad \forall\, v \in V_0, \\
2.\,\text{b)} & V_0\text{—bounded} : |a(u, v)| \le \mu_2 \|u\|\,\|v\| \quad \forall\, u, v \in V_0.
\end{array}
\right\} \tag{4.7}
$$

Here the constants μ_1 and μ_2 are supposed to be positive. Recall that $V_0 \subset V$ is usually an infinite dimensional, closed, nontrivial subspace of the Hilbert space V equipped with the scalar product $(\cdot, \cdot)$ and the corresponding norm $\|\cdot\|$. The standard assumptions (4.7) are sufficient for existence and uniqueness, as Lax–Milgram's theorem states:

Theorem 4.4 (Lax–Milgram).

Let us assume that our standard assumptions (4.7) are fulfilled. Then there exists a unique solution $u \in V_0$ of the variational problem (4.6).

Proof. Using the operator representation of (4.6) and Riesz isomorphism between V_0 and its dual space V_0^*, we can easily rewrite problem (4.6) as an operator fixed point problem in V_0. Then the proof of the results stated in the theorem immediately follows from Banach's fixed point theorem (see [22]). Moreover, Banach's fixed point theorem provides us with the fixed point iteration for approximating the solution together with a priori and a posteriori iteration error estimates. Later on we will use this iteration method to solve the Galerkin equations.   ☐

4.2   Galerkin finite element discretization

This section provides the basic details of the FEM. First we consider the Galerkin method for solving the abstract variational problem (4.1) introduced in section 4.1. Then we show that the FEM is nothing more than the Galerkin method with special basis functions. We focus on the practical aspects of the implementation of the FEM. To illustrate the main implementation steps, we consider our model problem CHIP introduced in section 4.1. The section is concluded with some introductory analysis of the Galerkin FEM and an overview of some iterative solvers derived from Banach's fixed point theorem. More information on FEM can be found in the standard monograph by P. Ciarlet [22] as well as in lecture notes on numerical methods for PDEs (e.g., [3, 11, 50]).

4.2.1 The Galerkin method

The Galerkin method looks for some approximation $u_h \in V_{gh} = g_h + V_{0h} \subset V_g$ to the exact solution $u \in V_g$ of the variational problem (4.1) in some finite dimensional counterpart V_{gh} of the linear manifold V_g such that u_h satisfies the variational equation (4.1) for all test functions v_h from the finite dimensional subspace $V_{0h} \subset V_0$ generating V_{gh}.

To be more specific, let us first introduce the finite dimensional space

$$V_h = \text{span}\left\{\varphi^{(i)} : i \in \overline{\omega}_h\right\} = \left\{v_h = \sum_{i \in \overline{\omega}_h} v^{(i)}\varphi^{(i)}\right\} = \text{span}\,\overline{\Phi} \subset V \qquad (4.8)$$

spanned by the (linearly independent) basis functions $\overline{\Phi} = [\varphi^{(i)} : i \in \overline{\omega}_h] = [\varphi_1, \ldots, \varphi_{\overline{N}_h}]$, where $\overline{\omega}_h$ is the index set providing the "numbering" of the basis functions. Due to the linear independence of the basis function, the dimension, $\dim V_h$, of the space V_h is given by $\dim V_h = |\overline{\omega}_h| = \overline{N}_h = N_h + \partial N_h < \infty$. After introducing the basic finite dimensional space V_h, we can define the finite dimensional linear manifold

$$V_{gh} = V_h \cap V_g = g_h + V_{0h} = \left\{v_h = \sum_{i \in \gamma_h} u_*^{(i)}\varphi^{(i)} + \sum_{i \in \omega_h} v^{(i)}\varphi^{(i)}\right\} \subset V_g \qquad (4.9)$$

and the corresponding finite dimensional test space

$$V_{0h} = V_h \cap V_0 = \left\{v_h = \sum_{i \in \omega_h} v^{(i)}\varphi^{(i)}\right\} = \text{span}\,\Phi \subset V_0, \qquad (4.10)$$

where $\dim V_{0h} = |\omega_h| = N_h$. Note that we implicitly assume that $V_h \cap V_g \neq \emptyset$ and that at least one element $g_h = \sum_{i \in \gamma_h} u_*^{(i)}\varphi^{(i)} \in V_g \cap V_h$ is available, where $\gamma_h := \overline{\omega}_h \setminus \omega_h$. For instance, the 1D linear basis functions are presented in the following picture:

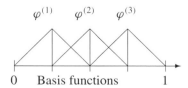

Once the basis functions are defined, it is easy to verify that the *Galerkin scheme* is equivalent to the *Galerkin system* of algebraic equations for defining the Galerkin coefficient vector, which is named the skeleton solution (SS), as illustrated in Fig. 4.2.

Before we analyze the Galerkin scheme (existence, uniqueness, and discretization error estimates) and discuss solvers for the system of equations

$$K_h \underline{u}_h = \underline{f}_h \qquad (4.11)$$

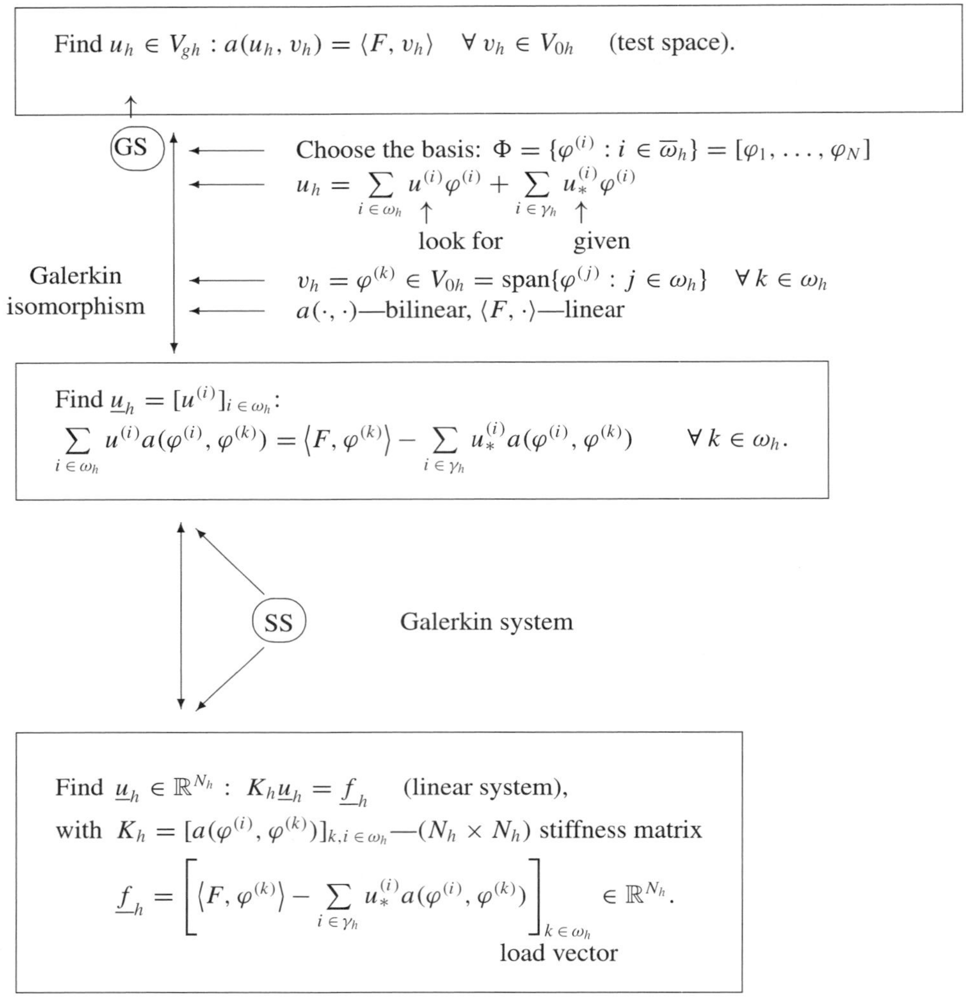

Find $u_h \in V_{gh} : a(u_h, v_h) = \langle F, v_h \rangle \quad \forall\, v_h \in V_{0h}$ (test space).

$\uparrow$

(GS) $\longleftarrow$ Choose the basis: $\Phi = \{\varphi^{(i)} : i \in \overline{\omega}_h\} = [\varphi_1, \ldots, \varphi_N]$

$\longleftarrow$ $u_h = \sum_{i \in \omega_h} u^{(i)}\varphi^{(i)} + \sum_{i \in \gamma_h} u_*^{(i)}\varphi^{(i)}$

     $\uparrow$ $\uparrow$
     look for   given

Galerkin $\longleftarrow$ $v_h = \varphi^{(k)} \in V_{0h} = \mathrm{span}\{\varphi^{(j)} : j \in \omega_h\} \quad \forall\, k \in \omega_h$
isomorphism $\longleftarrow$ $a(\cdot, \cdot)$—bilinear, $\langle F, \cdot \rangle$—linear

Find $\underline{u}_h = [u^{(i)}]_{i \in \omega_h}$:

$$\sum_{i \in \omega_h} u^{(i)} a(\varphi^{(i)}, \varphi^{(k)}) = \langle F, \varphi^{(k)} \rangle - \sum_{i \in \gamma_h} u_*^{(i)} a(\varphi^{(i)}, \varphi^{(k)}) \qquad \forall\, k \in \omega_h.$$

(SS)   Galerkin system

Find $\underline{u}_h \in \mathbb{R}^{N_h} : K_h \underline{u}_h = \underline{f}_h$ (linear system),

with $K_h = [a(\varphi^{(i)}, \varphi^{(k)})]_{k, i \in \omega_h}$—$(N_h \times N_h)$ stiffness matrix

$$\underline{f}_h = \left[\langle F, \varphi^{(k)} \rangle - \sum_{i \in \gamma_h} u_*^{(i)} a(\varphi^{(i)}, \varphi^{(k)}) \right]_{k \in \omega_h} \in \mathbb{R}^{N_h}.$$

load vector

Figure 4.2. *Galerkin scheme.*

arising from the Galerkin discretization, we introduce the FEM as a Galerkin method with special basis functions resulting in what is probably the most powerful method in the world for solving elliptic PDEs.

4.2.2   The simplest finite element schemes

The basic idea for overcoming the principal difficulties of the classical Galerkin method (i.e., the use of basis and test functions with global supports, e.g., polynomials) goes back to the paper [24] by R. Courant (1943), at least in a strong mathematical sense, and was recovered by engineers in the mid 1950s (see also the historical review [4] by I. Babuška): Use basis and test functions $\varphi^{(i)} = \varphi^{(i)}(x)$ with local supports, where the $\varphi^{(i)}$ can be defined element-wise by form functions.

   In order to construct such basis functions, the computational domain Ω is decomposed into "sufficiently small" subdomains δ_r, the so-called finite elements (e.g., triangles or quadrilaterals in two dimensions; tetrahedrons or hexahedrons in three dimensions). Courant defined his basis function $\varphi^{(i)}$ on some triangulation (Ω is decomposed into triangles) in such a way that it is continuous, linear on each triangle, 1 at the nodal point (vertex of the triangle) $x^{(i)}$ to which it belongs, and 0 at all the other nodal points of the triangulation; see Fig. 4.3. Due to its shape, Courant's basis function is also called the "hat function" or "chapeau function."

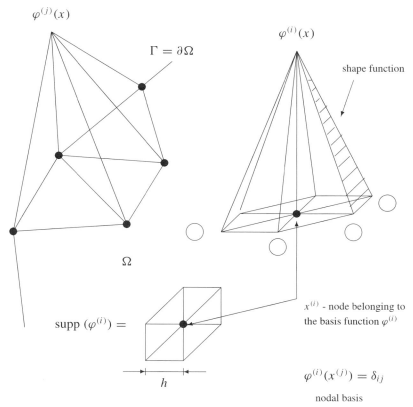

Figure 4.3. *Courant's basis function.*

   Using such basis functions, we search for the Galerkin solution (GS) of our variational problem in the form

$$u_h(x) = \sum_{i \in \bar{\omega}_h} u^{(i)} \varphi^{(i)}(x) \in V_h.$$

The piecewise linear finite element function u_h is obviously continuous, but not continuously differentiable across the edges of the elements. Due to the nodal basis property $\varphi^{(i)}(x^{(j)}) = \delta_{ij}$, the value $u_h(x^{(i)})$ of the finite element function u_h at the nodal point $x^{(i)}$ is equal to the coefficient $u^{(i)}$ for all nodal points $x^{(i)} \in \overline{\Omega}$, where δ_{ij} denotes Kronecker's symbol ($\delta_{ij} = 1$ for $i = j$ and $\delta_{ij} = 0$ for $i \neq j$). The use of such finite basis functions with local supports instead of basis functions with global supports is advantageous from various points of view:

1. There is great flexibility in satisfying the Dirichlet BCs.

2. The system matrix (= stiffness matrix) is sparse because of the locality of support of the basis and test functions; i.e.,

$$a(\varphi^{(i)}, \varphi^{(k)}) = 0 \text{ if } \operatorname{supp}(\varphi^{(i)}) \cap \operatorname{supp}(\varphi^{(k)}) = \emptyset.$$

3. The calculation of the coefficients of the stiffness matrix K_h and the load vector $\underline{f}_h$ is based on the quite efficient *element-wise* numerical integration technique.

4. It is easy to construct better and better finite element approximations u_h to the exact solution u of our variational problem by refining the finite elements even within an adaptive procedure ("limiting completeness" of the family of fine element spaces V_{0h}).

For the sake of simplicity of the presentation, we assume in the following that our computational domain Ω is plain and polygonally bounded. It is clear that such a computational domain can be decomposed into triangular finite elements δ_r exactly. Furthermore, we restrict ourselves to Courant's basis functions mentioned above (the linear triangular finite elements). Of course, this approach can easily be generalized to more complicated computational domains in two dimensions and three dimensions as well as to other finite elements (see [22, 73, 116]).

Triangulation (domain discretization)

The first step in the finite element Galerkin approach consists of the discretization of the computational domain—the *triangulation*. The triangulation $\mathcal{T}_h = \{\delta_r : r \in R_h := \{1, 2, \ldots, R_h\}\}$ is obtained by decomposing the computational domain Ω into triangles δ_r such that the following conditions are fulfilled:

1. $\overline{\Omega} = \bigcup_{r \in R_h} \overline{\delta}_r.$

2.   For all $r, r' \in R_h$, $r \neq r'$, the intersection $\overline{\delta}_r \cap \overline{\delta}_{r'}$ is given by

$$\overline{\delta}_r \cap \overline{\delta}_{r'} = \begin{cases} \emptyset & \text{or} \\ \text{a common edge} & \text{or} \\ \text{a common vertex.} \end{cases}$$

Among other things, the manner of the decomposition of Ω depends on the following:

1. the geometry of the domain (e.g., re-entrant corners on the boundary $\partial \Omega$ and micro-scales in the domain);

2. the input data of the elliptic BVP (e.g., jumping coefficients → interfaces, behavior of the right-hand side, mixed BCs → points where the type of the BCs is changing);

3. the accuracy imposed on the finite element solution (→ choice of the fineness of the mesh and/or the degree of the shape functions forming the finite element basis functions).

The following general hints should be taken into account by the triangulation procedure used:

1. Due to the second condition mentioned above, the triangulations in Fig. 4.4 are inadmissible.

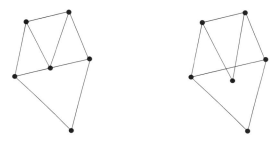

Figure 4.4. *Examples of inadmissible meshes.*

2. The mesh should be adapted to the change in the type of the BCs (see Fig. 4.5).

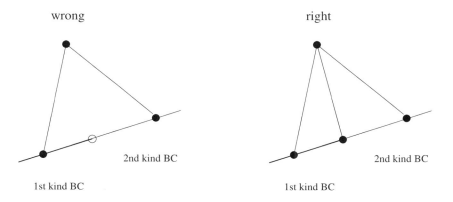

Figure 4.5. *Change in the type of the BCs.*

3. The interfaces should be captured by the triangulation (see Fig. 4.6).

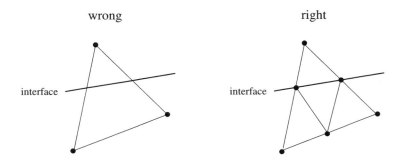

Figure 4.6. *Capturing interfaces by triangulation.*

4. The mesh should be refined appropriately in regions with a strongly changing gradient. This strongly changing gradient is caused by re-entrant corners on the boundary, corners in the interfaces, boundary points where the BCs change, etc.

5. Avoid triangles with too acute, or too obtuse, angles, especially triangles of the form shown in Fig. 4.7 (they cause a bad condition for the stiffness matrix and eventually bad approximation properties of the finite element solution).

Figure 4.7. *Triangle with an obtuse angle.*

The triangulation procedure (mesh generator) for plain, polygonally bounded computational domains Ω produces the coordinates of the vertices (or nodes) of the triangles. We construct both a *global* and a *local* numbering of the vertices and the *connectivity* of the vertices. This is called a discrete decomposition of the domain.

Global: Numbering of all nodes and elements:

$\overline{\omega}_h = \{1, 2, \ldots, \bar{N}_h\}$, $\bar{N}_h$—number of nodes,

$R_h = \{1, 2, \ldots, R_h\}$, R_h—number of elements.

Determination of the nodal coordinates:

$x^{(i)} = (x_{1,i}, x_{2,i})$, $i = 1, 2, \ldots, \bar{N}_h$.

Local: Anticlockwise numbering of the nodes in every triangle; see Fig. 4.8.

Connectivity: Determination of the connectivity between the local and global numberings of the nodes for all triangles $\delta^{(r)}$, $r \in R_h$:

$$r : \quad \alpha \leftrightarrow i = i(r, \alpha), \quad \alpha \in A^{(r)}, \ i \in \overline{\omega}_h.$$

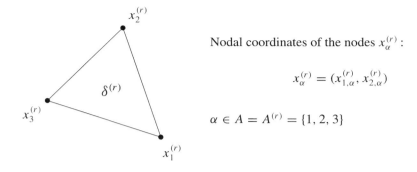

Nodal coordinates of the nodes $x_\alpha^{(r)}$:

$$x_\alpha^{(r)} = (x_{1,\alpha}^{(r)}, x_{2,\alpha}^{(r)})$$

$$\alpha \in A = A^{(r)} = \{1, 2, 3\}$$

Figure 4.8. *Local numbering of the nodes in a triangle.*

Therefore, a finite element code usually requires at least two arrays describing the triangulation, namely, the so-called finite element connectivity array $r : \alpha \leftrightarrow i$, $r \in R_h$, $\alpha \in A^{(r)} = A$, $i \in \overline{\omega}_h$, and the real array of all nodal coordinates $i : (x_{1,i}, x_{2,i})$. These two arrays are presented by Tables 4.1 and 4.2 for the triangulation of model problem CHIP shown in Fig. 4.9.

Element number	Global nodal numbers of the local nodes			Material property (MP)
	$\varphi_1^{(r)}$	$\varphi_2^{(r)}$	$\varphi_3^{(r)}$	
1	1	2	6	1
2	2	7	6	1
⋮	⋮	⋮	⋮	⋮
19	8	14	13	1
20	8	9	14	1
⋮	⋮	⋮	⋮	⋮
$R_h = 24$	12	13	17	2

Table 4.1. *Element connectivity table.*

i	1	2	3	$\cdots$	$\bar{N}_h = 21$
$x_{1,i}$	0.0	0.17	0.5	$\cdots$	0.65
$x_{2,i}$	0.0	0.0	0.0	$\cdots$	0.3

Table 4.2. *Table of the nodal coordinates.*

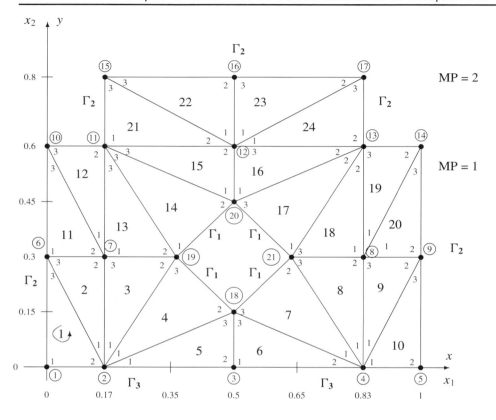

Figure 4.9. *Triangulation of the model problem CHIP.*

Further arrays for characterizing nodes, or elements, are required by many finite element codes, e.g., in order to assign MPs to the elements (see Table 4.1), or in order to describe the boundary $\partial\Omega$ and to code the BCs (see, e.g., [73]).

There are several methods for generating the mesh data from some geometrical description of the computational domain and from the PDE data. If it is possible to decompose the computational domain into a coarse mesh, as for our model problem CHIP, then one can simply edit the arrays described above by hand. More complicated 2D, or 3D, computational domains require the use of mesh generators that automatically produce meshes from the input data (see [39, 100, 73]). These hand-generated or automatically generated meshes can then be refined further via some a priori or a posteriori refinement procedures [109].

Definition of the finite element basis function

Once the mesh is generated, the finite element basis functions $\varphi^{(i)}$ can be locally defined via the element shape functions $\varphi_\alpha^{(r)}$ that are obtained from the master element shape functions by mapping.

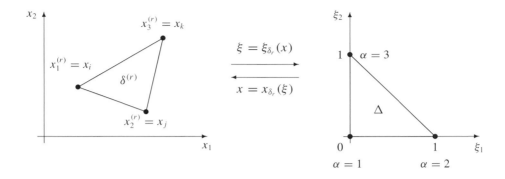

arbitrary triangle $\delta^{(r)} \in \mathcal{T}_h$ master triangle Δ

Figure 4.10. *Mapping between an arbitrary element and the master element.*

The affine linear transformation $x = J_{\delta^{(r)}} \xi + x^{(i)}$ is given by

$$\begin{pmatrix} x_1 \\ x_2 \end{pmatrix} = \begin{pmatrix} x_{1,j} - x_{1,i} & x_{1,k} - x_{1,i} \\ x_{2,j} - x_{2,i} & x_{2,k} - x_{2,i} \end{pmatrix} \begin{pmatrix} \xi_1 \\ \xi_2 \end{pmatrix} + \begin{pmatrix} x_{1,i} \\ x_{2,i} \end{pmatrix}. \tag{4.12}$$

This transformation maps the master element Δ (= unit triangle) onto an arbitrary triangle $\delta^{(r)}$ of our triangulation $\mathcal{T}_h$; see Fig. 4.10. The relation

$$|\det J_{\delta^{(r)}}| = 2 \operatorname{meas} \delta^{(r)}$$

immediately follows from the elementary calculations

$$\operatorname{meas} \delta^{(r)} = \int_{\delta^{(r)}} dx = \int_{\Delta} |\det J_{\delta^{(r)}}| \, d\xi = |\det J_{\delta^{(r)}}| \int_{\Delta} d\xi = \frac{1}{2} |\det J_{\delta^{(r)}}|,$$

where $\operatorname{meas} \delta^{(r)}$ denotes the area of the element $\delta^{(r)}$. The inverse mapping $\xi = J_{\delta^{(r)}}^{-1} (x - x_i)$ is given by the formulas

$$\begin{pmatrix} \xi_1 \\ \xi_2 \end{pmatrix} = \frac{1}{\det J_{\delta^{(r)}}} \begin{pmatrix} x_{2,k} - x_{2,i} & -(x_{1,k} - x_{1,i}) \\ -(x_{2,j} - x_{2,i}) & x_{1,j} - x_{1,i} \end{pmatrix} \begin{pmatrix} x_1 - x_{1,i} \\ x_2 - x_{2,i} \end{pmatrix}.$$

Now, using the element connectivity relation $r : \alpha \leftrightarrow i$ and the affine linear transformations, we can obviously represent the finite element basis functions $\varphi^{(i)}$ by the formula

$$\varphi^{(i)}(x) = \begin{cases} \varphi_\alpha^{(r)}(x), & x \in \bar{\delta}^{(r)}, \ r \in B_i \quad \text{(shape function)}, \\ 0 & \text{otherwise, i.e., } \forall \, x \in \overline{\Omega} \setminus \bigcup_{r \in B_i} \bar{\delta}^{(r)}, \end{cases} \tag{4.13}$$

where $B_i = \{r \in R_h : x^{(i)} \in \bar{\delta}^{(r)}\}$, $\varphi_\alpha^{(r)}(x) = \phi_\alpha(\xi_{\delta^{(r)}}(x))$, $x \in \bar{\delta}^{(r)}$, and ϕ_α denotes the master element shape function that corresponds to the element shape function $\varphi_\alpha^{(r)}$. For

Courant's triangle, we have three master element shape functions:

$$\left\{ \begin{array}{ccl} \phi_1 & = & 1 - \xi_1 - \xi_2, \\ \phi_2 & = & \xi_1, \\ \phi_3 & = & \xi_2, \end{array} \right. \tag{4.14}$$

which correspond to the nodes $(0, 0)$, $(1, 0)$, and $(0, 1)$, respectively.

Let us consider the triangulation of our model problem CHIP given in Fig. 4.9 as an example. If we choose, e.g., the node with the global number 8, then $B_8 = \{8, 9, 18, 19, 20\}$. The element-wise definition of the basis function $\varphi^{(8)}$ is now illustrated in Fig. 4.11.

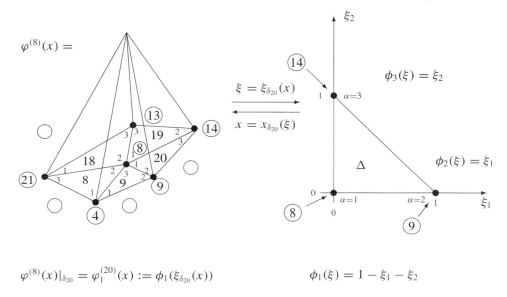

$$\varphi^{(8)}(x)|_{\delta_{20}} = \varphi_1^{(20)}(x) := \phi_1(\xi_{\delta_{20}}(x)) \qquad\qquad \phi_1(\xi) = 1 - \xi_1 - \xi_2$$

Figure 4.11. *Definition of the finite element basis function $\varphi^{(8)}$.*

Now it is clear that

$$\varphi^{(i)}(x_j) \;=\; \delta_{ij} \quad \text{for all } i, j \in \overline{\omega}_h \;(\longrightarrow \text{nodal basis}),$$

$$V_h \;=\; \left\{ v_h(x) \,:\, v_h(x) = \sum_{i \in \overline{\omega}_h} v^{(i)} \varphi^{(i)}(x) \right\} \subset V = H^1(\Omega),$$

$$V_{0h} \;=\; \left\{ v_h(x) \,:\, v_h(x) = \sum_{i \in \omega_h} v^{(i)} \varphi^{(i)}(x) \right\} \subset V_0,$$

$$V_{g_1 h} \;=\; \left\{ v_h(x) \,:\, v_h(x) = \sum_{i \in \omega_h} v^{(i)} \varphi^{(i)}(x) + \sum_{i \in \gamma_h} g_1(x_i) \varphi^{(i)}(x) \right\}.$$

The nodal basis relation $\varphi^{(i)}(x^{(j)}) = \delta_{ij}$ implies that $v_h(x^{(i)}) = v^{(i)}$. For the triangulation given in Fig. 4.9, we have

$$\overline{\omega}_h = \{1, 2, \ldots, 21\} \text{ and } \omega_h = \overline{\omega}_h \setminus \gamma_h, \ \gamma_h = \{18, 19, 20, 21\}.$$

Assembling the finite element equations

Assembling the finite element equations is a simple procedure, but is sometimes quite confusing the first time the process is seen. Strang and Fix [98] provide many details for a 1D BVP that the reader may find easier to follow than that below.

We are now going to describe the element-wise assembling of the system of finite element equations that is nothing more than the Galerkin system for our finite element basis functions (cf. section 4.2.1): Find $\underline{u}_h \in \mathbb{R}^{N_h}$ such that

$$K_h \underline{u}_h = \underline{f}_h, \tag{4.15}$$

with

$$K_h = [a(\varphi^{(i)}, \varphi^{(k)})]_{i,k \in \omega_h},$$

$$\underline{f}_h = \left[\langle F, \varphi^{(k)} \rangle - \sum_{i \in \gamma_h} u_*^{(i)} a(\varphi^{(i)}, \varphi^{(k)}) \right]_{k \in \omega_h}.$$

For our model problem CHIP, which we will consider shortly, the bilinear form $a(\cdot, \cdot)$ and the linear form $\langle F, \cdot \rangle$ are defined as follows (cf. again section 4.2.1):

$$a(\varphi^{(i)}, \varphi^{(k)}) = \int_\Omega \left[\lambda_1 \frac{\partial \varphi^{(i)}}{\partial x_1} \frac{\partial \varphi^{(k)}}{\partial x_1} + \lambda_2 \frac{\partial \varphi^{(i)}}{\partial x_2} \frac{\partial \varphi^{(k)}}{\partial x_2} \right] dx + \int_{\Gamma_3} \alpha \varphi^{(i)} \varphi^{(k)} ds,$$

$$\langle F, \varphi^{(k)} \rangle = \int_\Omega f(x) \varphi^{(k)}(x) dx + \int_{\Gamma_2} g_2(x) \varphi^{(k)}(x) ds + \int_{\Gamma_3} \alpha g_3 \varphi^{(k)} ds.$$

In the first step, we ignore the BCs and generate the extended matrix

$$\overline{K}_h = [\overline{a}(\varphi^{(i)}, \varphi^{(k)})]_{i,k \in \overline{\omega}_h} = \left[\int_\Omega \left[\lambda_1 \frac{\partial \varphi^{(i)}}{\partial x_1} \frac{\partial \varphi^{(k)}}{\partial x_1} + \lambda_2 \frac{\partial \varphi^{(i)}}{\partial x_2} \frac{\partial \varphi^{(k)}}{\partial x_2} \right] dx \right]_{i,k \in \overline{\omega}_h}$$

and the extended right-hand side

$$\underline{\overline{f}}_h = [\langle \overline{F}, \varphi^{(k)} \rangle]_{k \in \overline{\omega}_h} = \left[\int_\Omega f(x) \varphi^{(k)}(x) dx \right]_{k \in \overline{\omega}_h}.$$

The finite element technology for the element-wise generation of the matrix $\overline{K}_h$ and the right-hand side $\underline{\overline{f}}_h$ is based on the fundamental identities

$$(\overline{K}_h \underline{u}_h, \underline{v}_h) = \overline{a}(u_h, v_h) \quad \forall \ u_h \leftrightarrow \underline{u}_h, \ v_h \leftrightarrow \underline{v}_h, \ u_h, \ v_h \in V_h, \ \underline{u}_h, \ \underline{v}_h \in \mathbb{R}^{\overline{N}_h}$$

and

$$(\underline{\overline{f}}_h, \underline{v}_h) = \langle \overline{F}, v_h \rangle \quad \forall \ v_h \leftrightarrow \underline{v}_h, \ v_h \in V_h, \ \underline{v}_h \in \mathbb{R}^{\overline{N}_h},$$

respectively.

Generation of the element stiffness matrices

Inserting $u_h = \sum_{i\in\bar\omega_h} u^{(i)}\varphi^{(i)}(x)$ and $v_h = \sum_{k\in\bar\omega_h} v^{(k)}\varphi^{(k)}(x) \in V_h$ into the bilinear form $\bar a(\cdot,\cdot)$, we obtain the representation

$$\bar a(u_h, v_h) = \bar a\Big(\sum_{i\in\bar\omega_h} u^{(i)}\varphi^{(i)}(x), \sum_{k\in\bar\omega_h} v^{(k)}\varphi^{(k)}(x)\Big)$$

$$= \int_\Omega \Big[\lambda_1 \frac{\partial}{\partial x_1}\Big(\sum_{i\in\bar\omega_h} u^{(i)}\varphi^{(i)}(x)\Big)\frac{\partial}{\partial x_1}\Big(\sum_{k\in\bar\omega_h} v^{(k)}\varphi^{(k)}(x)\Big)$$

$$+ \lambda_2 \frac{\partial}{\partial x_2}\Big(\sum_{i\in\bar\omega_h} u^{(i)}\varphi^{(i)}(x)\Big)\frac{\partial}{\partial x_2}\Big(\sum_{k\in\bar\omega_h} v^{(k)}\varphi^{(k)}(x)\Big)\Big]\,dx$$

$$= \sum_{r\in R_h}\sum_{i\in\bar\omega_h}\sum_{k\in\bar\omega_h} u^{(i)}v^{(k)}\int_{\delta^{(r)}}\Big[\lambda_1\frac{\partial\varphi^{(i)}}{\partial x_1}\frac{\partial\varphi^{(k)}}{\partial x_1} + \lambda_2\frac{\partial\varphi^{(i)}}{\partial x_2}\frac{\partial\varphi^{(k)}}{\partial x_2}\Big]\,dx.$$

Introducing the index set $\bar\omega_h^{(r)} = \{j \; : \; x^{(j)} = (x_{1,j}, x_{2,j}) \in \bar\delta^{(r)}\}$ and taking into account

$$\int_{\delta^{(r)}}\Big[\lambda_1\frac{\partial\varphi^{(i)}}{\partial x_1}\frac{\partial\varphi^{(k)}}{\partial x_1} + \lambda_2\frac{\partial\varphi^{(i)}}{\partial x_2}\frac{\partial\varphi^{(k)}}{\partial x_2}\Big]\,dx = 0$$

for $x^{(i)} \notin \bar\delta^{(r)}$, or $x^{(k)} \notin \bar\delta^{(r)}$, we arrive at the representation

$$\bar a(u_h, v_h) = \sum_{r\in R_h}\sum_{i\in\bar\omega_h^{(r)}}\sum_{k\in\bar\omega_h^{(r)}} u^{(i)}v^{(k)}\int_{\delta^{(r)}}\Big[\lambda_1\frac{\partial\varphi^{(i)}}{\partial x_1}\frac{\partial\varphi^{(k)}}{\partial x_1} + \lambda_2\frac{\partial\varphi^{(i)}}{\partial x_2}\frac{\partial\varphi^{(k)}}{\partial x_2}\Big]\,dx.$$

Taking into account the correspondences $r : i \leftrightarrow \alpha$, $k \leftrightarrow \beta$, with $\alpha, \beta \in A^{(r)} = \{1, 2, 3\}$, we finally obtain (see also relation (4.13))

$$\bar a(u_h, v_h) = \sum_{r\in R_h}\sum_{\alpha\in A^{(r)}}\sum_{\beta\in A^{(r)}} u_\alpha^{(r)}v_\beta^{(r)}\int_{\delta^{(r)}}\Big[\lambda_1\frac{\partial\varphi_\alpha^{(r)}}{\partial x_1}\frac{\partial\varphi_\beta^{(r)}}{\partial x_1} + \lambda_2\frac{\partial\varphi_\alpha^{(r)}}{\partial x_2}\frac{\partial\varphi_\beta^{(r)}}{\partial x_2}\Big]\,dx$$

$$= \sum_{r\in R_h}(v_1^{(r)}\; v_2^{(r)}\; v_3^{(r)})K^{(r)}\begin{pmatrix} u_1^{(r)} \\ u_2^{(r)} \\ u_3^{(r)} \end{pmatrix},$$

where the element stiffness matrices $K^{(r)}$ are defined as follows:

$$K^{(r)} = \Big[\int_{\delta^{(r)}}\Big[\lambda_1\frac{\partial\varphi_\alpha^{(r)}}{\partial x_1}\frac{\partial\varphi_\beta^{(r)}}{\partial x_1} + \lambda_2\frac{\partial\varphi_\alpha^{(r)}}{\partial x_2}\frac{\partial\varphi_\beta^{(r)}}{\partial x_2}\Big]\,dx\Big]_{\beta,\alpha=1}^3.$$

The coefficients of the element stiffness matrices $K^{(r)}$ will be calculated by transforming the integrals over $\delta^{(r)}$ into integrals over the master element $\Delta = \{(\xi_1, \xi_2) \; : \; 0 \le \xi_1 \le 1, \; 0 \le \xi_2 \le 1, \; \xi_1 + \xi_2 \le 1\}$. The transformation $x_{\delta^{(r)}} = x_{\delta^{(r)}}(\xi)$ of the master element onto the finite element $\delta^{(r)}$ is explicitly given by the formula $x = J_{\delta^{(r)}}\xi + x^{(i)}$; i.e.,

$$\begin{pmatrix} x_1 \\ x_2 \end{pmatrix} = \begin{pmatrix} x_{1,j} - x_{1,i} & x_{1,k} - x_{1,i} \\ x_{2,j} - x_{2,i} & x_{2,k} - x_{2,i} \end{pmatrix}\begin{pmatrix} \xi_1 \\ \xi_2 \end{pmatrix} + \begin{pmatrix} x_{1,i} \\ x_{2,i} \end{pmatrix},$$

where $(x_{1,i}, x_{2,i})$, $(x_{1,j}, x_{2,j})$, $(x_{1,k}, x_{2,k})$ denote the vertices of the triangle $\delta^{(r)}$ and the correspondence $i \leftrightarrow 1$, $j \leftrightarrow 2$, $k \leftrightarrow 3$ between the global and local nodal numberings is taken from the element connectivity table $r : i \leftrightarrow \alpha$. The inverse mapping $\xi_{\delta^{(r)}} = \xi_{\delta^{(r)}}(x)$ is obviously given by the formula

$$
\begin{pmatrix} \xi_1 \\ \xi_2 \end{pmatrix} = J_{\delta^{(r)}}^{-1} \begin{pmatrix} x_1 - x_{1,i} \\ x_2 - x_{2,i} \end{pmatrix}
$$

$$
= \frac{1}{\det J_{\delta^{(r)}}} \begin{pmatrix} x_{2,k} - x_{2,i} & -(x_{1,k} - x_{1,i}) \\ -(x_{2,j} - x_{2,i}) & x_{1,j} - x_{1,i} \end{pmatrix} \begin{pmatrix} x_1 - x_{1,i} \\ x_2 - x_{2,i} \end{pmatrix}
$$

$$
= \begin{pmatrix} \frac{\partial \xi_1}{\partial x_1} & \frac{\partial \xi_1}{\partial x_2} \\ \frac{\partial \xi_2}{\partial x_1} & \frac{\partial \xi_2}{\partial x_2} \end{pmatrix} \begin{pmatrix} x_1 - x_{1,i} \\ x_2 - x_{2,i} \end{pmatrix}.
$$

The transformation formulas

$$
\frac{\partial}{\partial x_1} = \frac{\partial}{\partial \xi_1} \frac{\partial \xi_1}{\partial x_1} + \frac{\partial}{\partial \xi_2} \frac{\partial \xi_2}{\partial x_1} \quad \text{and} \quad \frac{\partial}{\partial x_2} = \frac{\partial}{\partial \xi_1} \frac{\partial \xi_1}{\partial x_2} + \frac{\partial}{\partial \xi_2} \frac{\partial \xi_2}{\partial x_2}
$$

for the partial derivatives give the relations

$$
\frac{\partial}{\partial x_1} = \frac{1}{\det J_{\delta^{(r)}}} \left[(x_{2,k} - x_{2,i}) \frac{\partial}{\partial \xi_1} - (x_{2,j} - x_{2,i}) \frac{\partial}{\partial \xi_2} \right],
$$

$$
\frac{\partial}{\partial x_2} = \frac{1}{\det J_{\delta^{(r)}}} \left[-(x_{1,k} - x_{1,i}) \frac{\partial}{\partial \xi_1} + (x_{1,j} - x_{1,i}) \frac{\partial}{\partial \xi_2} \right].
$$

Using these transformation formulas and the relation (4.13), we obtain the formulas for the coefficients $K_{\beta\alpha}^{(r)}$ of the element stiffness matrix $K^{(r)}$:

$$
\begin{aligned}
K_{\beta\alpha}^{(r)} &= \int_{\delta^{(r)}} \left[\lambda_1 \frac{\partial \varphi_\alpha^{(r)}}{\partial x_1} \frac{\partial \varphi_\beta^{(r)}}{\partial x_1} + \lambda_2 \frac{\partial \varphi_\alpha^{(r)}}{\partial x_2} \frac{\partial \varphi_\beta^{(r)}}{\partial x_2} \right] dx \\
&= \int_\Delta \left[\lambda_1(x_{\delta^{(r)}}(\xi)) \left(\frac{\partial \varphi_\alpha^{(r)}(x_{\delta^{(r)}}(\xi))}{\partial \xi_1} \frac{\partial \xi_1}{\partial x_1} + \frac{\partial \varphi_\alpha^{(r)}(x_{\delta^{(r)}}(\xi))}{\partial \xi_2} \frac{\partial \xi_2}{\partial x_1} \right) \right. \\
&\quad \cdot \left(\frac{\partial \varphi_\beta^{(r)}(x_{\delta^{(r)}}(\xi))}{\partial \xi_1} \frac{\partial \xi_1}{\partial x_1} + \frac{\partial \varphi_\beta^{(r)}(x_{\delta^{(r)}}(\xi))}{\partial \xi_2} \frac{\partial \xi_2}{\partial x_1} \right) \\
&\quad + \lambda_2(x_{\delta^{(r)}}(\xi)) \left(\frac{\partial \varphi_\alpha^{(r)}(x_{\delta^{(r)}}(\xi))}{\partial \xi_1} \frac{\partial \xi_1}{\partial x_2} + \frac{\partial \varphi_\alpha^{(r)}(x_{\delta^{(r)}}(\xi))}{\partial \xi_2} \frac{\partial \xi_2}{\partial x_2} \right) \\
&\quad \left. \cdot \left(\frac{\partial \varphi_\beta^{(r)}(x_{\delta^{(r)}}(\xi))}{\partial \xi_1} \frac{\partial \xi_1}{\partial x_2} + \frac{\partial \varphi_\beta^{(r)}(x_{\delta^{(r)}}(\xi))}{\partial \xi_2} \frac{\partial \xi_2}{\partial x_2} \right) \right] |\det J_{\delta^{(r)}}| \, d\xi \\
&= \int_\Delta |\det J_{\delta^{(r)}}| \left[\lambda_1(x_{\delta^{(r)}}(\xi)) \left(\frac{\partial \phi_\alpha(\xi)}{\partial \xi_1} \frac{\partial \xi_1}{\partial x_1} + \frac{\partial \phi_\alpha(\xi)}{\partial \xi_2} \frac{\partial \xi_2}{\partial x_1} \right) \left(\frac{\partial \phi_\beta(\xi)}{\partial \xi_1} \frac{\partial \xi_1}{\partial x_1} + \frac{\partial \phi_\beta(\xi)}{\partial \xi_2} \frac{\partial \xi_2}{\partial x_1} \right) \right. \\
&\quad \left. + \lambda_2(x_{\delta^{(r)}}(\xi)) \left(\frac{\partial \phi_\alpha(\xi)}{\partial \xi_1} \frac{\partial \xi_1}{\partial x_2} + \frac{\partial \phi_\alpha(\xi)}{\partial \xi_2} \frac{\partial \xi_2}{\partial x_2} \right) \left(\frac{\partial \phi_\beta(\xi)}{\partial \xi_1} \frac{\partial \xi_1}{\partial x_2} + \frac{\partial \phi_\beta(\xi)}{\partial \xi_2} \frac{\partial \xi_2}{\partial x_2} \right) \right] d\xi.
\end{aligned}
$$

In the case of linear shape functions (4.14):

$$\phi_1(\xi) = 1 - \xi_1 - \xi_2, \qquad \frac{\partial \phi_1}{\partial \xi_1} = -1, \qquad \frac{\partial \phi_1}{\partial \xi_2} = -1,$$

$$\phi_2(\xi) = \xi_1, \qquad \frac{\partial \phi_2}{\partial \xi_1} = 1, \qquad \frac{\partial \phi_2}{\partial \xi_2} = 0,$$

$$\phi_3(\xi) = \xi_2, \qquad \frac{\partial \phi_3}{\partial \xi_1} = 0, \qquad \frac{\partial \phi_3}{\partial \xi_2} = 1,$$

we derive from the above formulas the explicit relations

$$
\begin{aligned}
K_{11}^{(r)} &= \frac{1}{|\det J_{\delta^{(r)}}|} \int_\Delta [\lambda_1(x_{\delta^{(r)}}(\xi))(-(x_{2,k} - x_{2,i}) + (x_{2,j} - x_{2,i}))^2 \\
&\qquad + \lambda_2(x_{\delta^{(r)}}(\xi))((x_{1,k} - x_{1,i}) - (x_{1,j} - x_{1,i}))^2] \, d\xi \\
&= \frac{1}{|\det J_{\delta^{(r)}}|} \left((x_{2,j} - x_{2,k})^2 \int_\Delta \lambda_1(x_{\delta^{(r)}}(\xi)) \, d\xi + (x_{1,k} - x_{1,j})^2 \int_\Delta \lambda_2(x_{\delta^{(r)}}(\xi)) \, d\xi \right), \\
K_{22}^{(r)} &= \frac{1}{|\det J_{\delta^{(r)}}|} \int_\Delta [\lambda_1(x_{\delta^{(r)}}(\xi))(x_{2,k} - x_{2,i})^2 + \lambda_2(x_{\delta^{(r)}}(\xi))(-(x_{1,k} - x_{1,i}))^2] \, d\xi \\
&= \frac{1}{|\det J_{\delta^{(r)}}|} \left((x_{2,k} - x_{2,i})^2 \int_\Delta \lambda_1(x_{\delta^{(r)}}(\xi)) \, d\xi + (x_{1,i} - x_{1,k})^2 \int_\Delta \lambda_2(x_{\delta^{(r)}}(\xi)) \, d\xi \right), \\
K_{33}^{(r)} &= \frac{1}{|\det J_{\delta^{(r)}}|} \int_\Delta [\lambda_1(x_{\delta^{(r)}}(\xi))(-(x_{2,j} - x_{2,i}))^2 + \lambda_2(x_{\delta^{(r)}}(\xi))(x_{1,j} - x_{1,i})^2] \, d\xi \\
&= \frac{1}{|\det J_{\delta^{(r)}}|} \left((x_{2,i} - x_{2,j})^2 \int_\Delta \lambda_1(x_{\delta^{(r)}}(\xi)) \, d\xi + (x_{1,j} - x_{1,i})^2 \int_\Delta \lambda_2(x_{\delta^{(r)}}(\xi)) \, d\xi \right), \\
K_{12}^{(r)} &= \frac{1}{|\det J_{\delta^{(r)}}|} \int_\Delta [\lambda_1(x_{\delta^{(r)}}(\xi))(x_{2,k} - x_{2,i})(-(x_{2,k} - x_{2,i}) + (x_{2,j} - x_{2,i})) \\
&\qquad + \lambda_2(x_{\delta^{(r)}}(\xi))(-(x_{1,k} - x_{1,i})((x_{1,k} - x_{1,i}) - (x_{1,j} - x_{1,i})))] \, d\xi \\
&= \frac{1}{|\det J_{\delta^{(r)}}|} \left((x_{2,k} - x_{2,i})(x_{2,j} - x_{2,k}) \int_\Delta \lambda_1(x_{\delta^{(r)}}(\xi)) \, d\xi \right. \\
&\qquad \left. + (x_{1,i} - x_{1,k})(x_{1,k} - x_{1,j}) \int_\Delta \lambda_2(x_{\delta^{(r)}}(\xi)) \, d\xi \right), \\
K_{13}^{(r)} &= \frac{1}{|\det J_{\delta^{(r)}}|} \int_\Delta [\lambda_1(x_{\delta^{(r)}}(\xi))(-(x_{2,j} - x_{2,i})(-(x_{2,k} - x_{2,i}) + (x_{2,j} - x_{2,i}))) \\
&\qquad + \lambda_2(x_{\delta^{(r)}}(\xi))(x_{1,j} - x_{1,i})((x_{1,k} - x_{1,i}) - (x_{1,j} - x_{1,i}))] \, d\xi \\
&= \frac{1}{|\det J_{\delta^{(r)}}|} \left((x_{2,i} - x_{2,j})(x_{2,j} - x_{2,k}) \int_\Delta \lambda_1(x_{\delta^{(r)}}(\xi)) \, d\xi \right. \\
&\qquad \left. + (x_{1,j} - x_{1,i})(x_{1,k} - x_{1,j}) \int_\Delta \lambda_2(x_{\delta^{(r)}}(\xi)) \, d\xi \right), \\
K_{23}^{(r)} &= \frac{1}{|\det J_{\delta^{(r)}}|} \int_\Delta [\lambda_1(x_{\delta^{(r)}}(\xi))(-(x_{2,j} - x_{2,i})(x_{2,k} - x_{2,i})) \\
&\qquad + \lambda_2(x_{\delta^{(r)}}(\xi))(x_{1,j} - x_{1,i})(-(x_{1,k} - x_{1,i}))] \, d\xi \\
&= \frac{1}{|\det J_{\delta^{(r)}}|} \left((x_{2,i} - x_{2,j})(x_{2,k} - x_{2,i}) \int_\Delta \lambda_1(x_{\delta^{(r)}}(\xi)) \, d\xi \right. \\
&\qquad \left. + (x_{1,j} - x_{1,i})(x_{1,i} - x_{1,k}) \int_\Delta \lambda_2(x_{\delta^{(r)}}(\xi)) \, d\xi \right).
\end{aligned}
$$

These formulas can be implemented easily if the integrals $\int_\Delta \lambda_1 \, d\xi$ and $\int_\Delta \lambda_2 \, d\xi$ are easy to calculate. If λ_1 and λ_2 are constant on the elements, then $\int_\Delta \lambda_1 \, d\xi = \frac{1}{2}\lambda_1$ and $\int_\Delta \lambda_2 \, d\xi = \frac{1}{2}\lambda_2$. Otherwise, we use the midpoint rule

$$\int_\Delta w(\xi) \, d\xi \approx \frac{1}{2} w(\xi_s), \qquad \xi_s = \left(\frac{1}{3}, \frac{1}{3}\right).$$

Therefore,

$$\int_\Delta \lambda_1(x_{\delta^{(r)}}(\xi)) \, d\xi \approx \frac{1}{2}\lambda_1(x_{\delta^{(r)}}(\xi_s)), \qquad \xi_s = \left(\frac{1}{3}, \frac{1}{3}\right),$$

and

$$\int_\Delta \lambda_2(x_{\delta^{(r)}}(\xi)) \, d\xi \approx \frac{1}{2}\lambda_2(x_{\delta^{(r)}}(\xi_s)), \qquad \xi_s = \left(\frac{1}{3}, \frac{1}{3}\right).$$

We note that the midpoint rule exactly integrates first-order polynomials, i.e., polynomials of the form $\varphi(\xi) = a_0 + a_1\xi_1 + a_2\xi_2$. Setting $k_1 = \frac{1}{2}\lambda_1$, or $k_1 = \frac{1}{2}\lambda_1(x_{\delta^{(r)}}(\xi_s))$, and $k_2 = \frac{1}{2}\lambda_2$, or $k_2 = \frac{1}{2}\lambda_2(x_{\delta^{(r)}}(\xi_s))$, we can rewrite the element stiffness matrix $K^{(r)}$ in the form

$$K^{(r)} = \left(K^{(r)}\right)^T = \frac{1}{|\det J_{\delta^{(r)}}|}$$

$$\begin{pmatrix} k_1(x_{2,j} - x_{2,k})^2 & k_1(x_{2,k} - x_{2,i})(x_{2,j} - x_{2,k}) & k_1(x_{2,i} - x_{2,j})(x_{2,j} - x_{2,k}) \\ +k_2(x_{1,k} - x_{1,j})^2 & +k_2(x_{1,i} - x_{1,k})(x_{1,k} - x_{1,j}) & +k_2(x_{1,j} - x_{1,i})(x_{1,k} - x_{1,j}) \\ \\ & k_1(x_{2,k} - x_{2,i})^2 & k_1(x_{2,i} - x_{2,j})(x_{2,k} - x_{2,i}) \\ & +k_2(x_{1,i} - x_{1,k})^2 & +k_2(x_{1,j} - x_{1,i})(x_{1,i} - x_{1,k}) \\ \\ & & k_1(x_{2,i} - x_{2,j})^2 \\ & & +k_2(x_{1,j} - x_{1,i})^2 \end{pmatrix},$$

where we again assume the correspondence $i \leftrightarrow 1$, $j \leftrightarrow 2$, $k \leftrightarrow 3$ between the global and the local nodal numbering, as in Fig. 4.11.

We note that we have only used the coefficients λ_i of the PDE, the mapping $x = J_{\delta^{(r)}}\xi + x^{(i)}$, and the shape functions ϕ_α over the master element for calculating the coefficients $K_{\alpha\beta}^{(r)}$ of the element stiffness matrix $K^{(r)}$. The explicit form of the shape functions over the finite element $\delta^{(r)}$ is not required. This is also valid for higher-order shape functions.

Generation of the element load vectors

The element load vectors $\underline{f}^{(r)}$ will be generated in the same way as we have just generated the element stiffness matrix $K^{(r)}$, i.e., based on the relation

$$(\underline{\bar{f}}_h, \underline{v}_h) = \langle \bar{F}, v_h \rangle = \int_\Omega f(x) v_h(x) \, dx = \int_\Omega f(x) \sum_{k \in \overline{\omega}_h} v^{(k)} \varphi^{(k)}(x) \, dx$$

$$= \sum_{r \in R_h} \sum_{k \in \overline{\omega}_h} v^{(k)} \int_{\delta^{(r)}} f(x) \varphi^{(k)}(x) \, dx.$$

Introducing again the index set $\overline{\omega}_h^{(r)} = \{k \ : \ x^{(k)} = (x_{1,k}, x_{2,k}) \in \overline{\delta}^{(r)}\}$ and taking into account that $\varphi^{(k)}(x)|_{\delta^{(r)}} = 0$ if $x^{(k)} \notin \overline{\delta}^{(r)}$, we obtain the representation

$$(\bar{f}_h, \underline{v}_h) = \sum_{r \in R_h} \sum_{k \in \overline{\omega}_h^{(r)}} v^{(k)} \int_{\delta^{(r)}} f(x) \varphi^{(k)}(x) \, dx.$$

This can be rewritten in the form

$$(\bar{f}_h, \underline{v}_h) = \sum_{r \in R_h} \sum_{\beta \in A^{(r)}} v_\beta^{(r)} \int_{\delta^{(r)}} f(x) \varphi_\beta^{(r)}(x) \, dx = \sum_{r \in R_h} (v_1^{(r)} \ v_2^{(r)} \ v_3^{(r)}) \underline{f}^{(r)},$$

with the element connectivity relation $r \ : \ k \ \leftrightarrow \ \beta$, where the element load vectors $\underline{f}^{(r)}$ are defined by the relation

$$\underline{f}^{(r)} = \left[\int_{\delta^{(r)}} f(x) \varphi_\beta^{(r)}(x) \, dx \right]_{\beta=1}^3.$$

We again transform the integrals over $\delta^{(r)}$ into integrals over the master element

$$\int_{\delta^{(r)}} f(x) \varphi_\beta^{(r)}(x) \, dx = \int_\Delta f(x_{\delta^{(r)}}(\xi)) \varphi_\beta^{(r)}(x_{\delta^{(r)}}(\xi)) |\det J_{\delta^{(r)}}| \, d\xi$$

$$= \int_\Delta f(x_{\delta^{(r)}}(\xi)) \phi_\beta(\xi) |\det J_{\delta^{(r)}}| \, d\xi$$

that can be approximated by the midpoint rule

$$\int_\Delta f(x_{\delta^{(r)}}(\xi)) \varphi_\beta(\xi) |\det J_{\delta^{(r)}}| \, d\xi \approx \frac{1}{2} f(x_{\delta^{(r)}}(\xi_s)) \varphi_\beta(\xi_s) |\det J_{\delta^{(r)}}|,$$

with $\xi_s = (1/3, 1/3)$.

Assembling the element stiffness matrices and the element load vectors

To assemble the element stiffness matrices $K^{(r)}$ and the element load vectors $\underline{f}^{(r)}$ into the global stiffness matrix $\bar{K}_h$ and the global load vector $\bar{f}_h$, we only need the element connectivity table $r \ : \ i \ \leftrightarrow \ \alpha$, which relates the local nodal numbers to global ones. We recall that the element connectivity table is generated by the mesh generator.

   For our model problem CHIP, the element connectivity table belonging to the mesh shown in Fig. 4.9 is given in Table 4.1. Using this information, we now assemble the corresponding global stiffness matrix $\bar{K}_h$. First we set all coefficients of $\bar{K}_h$ to zero. Then we calculate the element stiffness matrices $K^{(r)}$ one by one and immediately add them to the global stiffness matrix $\bar{K}_h$ (loop over all elements). After adding $K^{(1)}$ to $\bar{K}_h$, we obtain the matrix in Fig. 4.12. The incorporation of $K^{(2)}$ into $\bar{K}_h$ yields the matrix in Fig. 4.13.

   After finishing this assembly process, we obtain the global stiffness matrix $\bar{K}_h$ without considering the BCs.

$$
\begin{array}{c}
\begin{array}{cccccccccccc}
 & 1 & 2 & 3 & \cdots & 5 & 6 & 7 & 8 & \cdots & 19 & 20 & 21
\end{array} \\
\begin{array}{c}
1 \\ 2 \\ 3 \\ \vdots \\ 5 \\ 6 \\ 7 \\ 8 \\ \vdots \\ 19 \\ 20 \\ 21
\end{array}
\left(
\begin{array}{cccccccccccc}
K_{11}^{(1)} & K_{12}^{(1)} & 0 & \cdots & 0 & K_{13}^{(1)} & 0 & 0 & \cdots & 0 & 0 & 0 \\
K_{21}^{(1)} & K_{22}^{(1)} & 0 & \cdots & 0 & K_{23}^{(1)} & 0 & 0 & \cdots & 0 & 0 & 0 \\
0 & 0 & 0 & \cdots & 0 & 0 & 0 & 0 & \cdots & 0 & 0 & 0 \\
\vdots & \vdots & \vdots & \ddots & \vdots & \vdots & \vdots & \vdots & \ddots & \vdots & \vdots & \vdots \\
0 & 0 & 0 & \cdots & 0 & 0 & 0 & 0 & \cdots & 0 & 0 & 0 \\
K_{31}^{(1)} & K_{32}^{(1)} & 0 & \cdots & 0 & K_{33}^{(1)} & 0 & 0 & \cdots & 0 & 0 & 0 \\
0 & 0 & 0 & \cdots & 0 & 0 & 0 & 0 & \cdots & 0 & 0 & 0 \\
0 & 0 & 0 & \cdots & 0 & 0 & 0 & 0 & \cdots & 0 & 0 & 0 \\
\vdots & \vdots & \vdots & \ddots & \vdots & \vdots & \vdots & \vdots & \ddots & \vdots & \vdots & \vdots \\
0 & 0 & 0 & \cdots & 0 & 0 & 0 & 0 & \cdots & 0 & 0 & 0 \\
0 & 0 & 0 & \cdots & 0 & 0 & 0 & 0 & \cdots & 0 & 0 & 0 \\
0 & 0 & 0 & \cdots & 0 & 0 & 0 & 0 & \cdots & 0 & 0 & 0
\end{array}
\right)
\end{array}
$$

Figure 4.12. *One element matrix is assembled into the global matrix.*

$$
\begin{array}{c}
\begin{array}{cccccccccccc}
 & 1 & 2 & 3 & \cdots & 5 & 6 & 7 & 8 & \cdots & 19 & 20 & 21
\end{array} \\
\begin{array}{c}
1 \\ 2 \\ 3 \\ \vdots \\ 5 \\ 6 \\ 7 \\ 8 \\ \vdots \\ 19 \\ 20 \\ 21
\end{array}
\left(
\begin{array}{cccccccccccc}
K_{11}^{(1)} & K_{12}^{(1)} & 0 & \cdots & 0 & K_{13}^{(1)} & 0 & 0 & \cdots & 0 & 0 & 0 \\
K_{21}^{(1)} & K_{22}^{(1)}+K_{11}^{(2)} & 0 & \cdots & 0 & K_{23}^{(1)}+K_{13}^{(2)} & K_{12}^{(2)} & 0 & \cdots & 0 & 0 & 0 \\
0 & 0 & 0 & \cdots & 0 & 0 & 0 & 0 & \cdots & 0 & 0 & 0 \\
\vdots & \vdots & \vdots & \ddots & \vdots & \vdots & \vdots & \vdots & \ddots & \vdots & \vdots & \vdots \\
0 & 0 & 0 & \cdots & 0 & 0 & 0 & 0 & \cdots & 0 & 0 & 0 \\
K_{31}^{(1)} & K_{32}^{(1)}+K_{31}^{(2)} & 0 & \cdots & 0 & K_{33}^{(1)}+K_{33}^{(2)} & K_{32}^{(2)} & 0 & \cdots & 0 & 0 & 0 \\
0 & K_{21}^{(2)} & 0 & \cdots & 0 & K_{23}^{(2)} & K_{22}^{(2)} & 0 & \cdots & 0 & 0 & 0 \\
0 & 0 & 0 & \cdots & 0 & 0 & 0 & 0 & \cdots & 0 & 0 & 0 \\
\vdots & \vdots & \vdots & \ddots & \vdots & \vdots & \vdots & \vdots & \ddots & \vdots & \vdots & \vdots \\
0 & 0 & 0 & \cdots & 0 & 0 & 0 & 0 & \cdots & 0 & 0 & 0 \\
0 & 0 & 0 & \cdots & 0 & 0 & 0 & 0 & \cdots & 0 & 0 & 0 \\
0 & 0 & 0 & \cdots & 0 & 0 & 0 & 0 & \cdots & 0 & 0 & 0
\end{array}
\right)
\end{array}
$$

Figure 4.13. *Two element matrices are added into the global matrix.*

The assembling of the right-hand side $\underline{\bar{f}}_h$ follows the same algorithm; i.e., we first set all coefficients of the vector $\underline{\bar{f}}_h$ to zero, then we calculate the element load vector $\underline{f}^{(1)}$ and add it to $\underline{\bar{f}}_h$, obtaining

$$\underline{\bar{f}}_h = \begin{pmatrix} \underset{1}{f_1^{(1)}} & \underset{2}{f_2^{(1)}} & \underset{3}{0} & \underset{\cdots}{\cdots} & \underset{5}{0} & \underset{6}{f_3^{(1)}} & \underset{7}{0} & \underset{8}{0} & \underset{\cdots}{\cdots} & \underset{19}{0} & \underset{20}{0} & \underset{21}{0} \end{pmatrix}^T .$$

After adding $\underline{f}^{(2)}$, we have

$$\underline{\bar{f}}_h = \begin{pmatrix} \underset{1}{f_1^{(1)}} & \underset{2}{f_2^{(1)} + f_1^{(2)}} & \underset{3}{0} & \underset{\cdots}{\cdots} & \underset{5}{0} & \underset{6}{f_3^{(1)} + f_3^{(2)}} & \underset{7}{f_2^{(2)}} & \underset{8}{0} & \underset{\cdots}{\cdots} & \underset{19}{0} & \underset{20}{0} & \underset{21}{0} \end{pmatrix}^T .$$

Repeating this assembly process for all $r = 3, 4, \ldots, R_h$ (looping over all elements), we arrive at the global load vector $\underline{\bar{f}}_h$ which does not take into account the BCs. For instance, the coefficient $\bar{f}^{(8)}$ is composed as follows:

$$\bar{f}^{(8)} = f_2^{(8)} + f_3^{(9)} + f_1^{(18)} + f_1^{(19)} + f_1^{(20)}.$$

We note that the matrix $\bar{K}_h$ and the right-hand side $\underline{\bar{f}}_h$ can also be represented by the use of the so-called element connectivity matrices $C^{(r)}$ defined by the relation

$$C_{ij}^{(r)} = \begin{cases} 1 & \text{if } i \text{ is a global nodal number of some vertex of the triangle } \delta^{(r)} \\ & \text{and } j \text{ is the corresponding local nodal number,} \\ 0 & \text{otherwise,} \end{cases}$$

with $i = 1, 2, \ldots, \bar{N}_h$ and $j = 1, 2, 3$. Therefore, we have the representations

$$\bar{K}_h = \sum_{r \in R_h} C^{(r)} K^{(r)} (C^{(r)})^T \quad \text{and} \quad \underline{\bar{f}}_h = \sum_{r \in R_h} C^{(r)} \underline{f}^{(r)}.$$

For instance, the matrix $(C^{(1)})^T$ has the form

$$(C^{(1)})^T = \begin{pmatrix} \underset{1}{1} & \underset{2}{0} & \underset{3}{0} & \underset{\cdots}{\cdots} & \underset{5}{0} & \underset{6}{0} & \underset{7}{0} & \underset{8}{0} & \underset{\cdots}{\cdots} & \underset{19}{0} & \underset{20}{0} & \underset{21}{0} \\ 0 & 1 & 0 & \cdots & 0 & 0 & 0 & 0 & \cdots & 0 & 0 & 0 \\ 0 & 0 & 0 & \cdots & 0 & 1 & 0 & 0 & \cdots & 0 & 0 & 0 \end{pmatrix}.$$

This immediately follows from the element connectivity in Table 4.1.

Up to now, we have not taken into account the contributions from the Neumann (2nd-kind) and Robin (3rd-kind) BCs:

$$\int_{\Gamma_2} g_2(x) v_h(x)\, ds , \quad \int_{\Gamma_3} \alpha(x) g_3(x) v_h(x)\, ds , \quad \text{and} \quad \int_{\Gamma_3} \alpha(x) u_h(x) v_h(x)\, ds .$$

The same holds for the Dirichlet (1st-kind) BCs which we also have to integrate into the global stiffness matrix and the global load vector.

Including 2nd- and 3rd-kind BCs

Recall that we assumed that our computational domain Ω is polygonally bounded. The conditions imposed on the mesh generation above imply that

$$\overline{\Gamma}_2 = \bigcup_{e_2 \in E_h^{(2)}} \overline{\Gamma}_2^{(e_2)} \qquad \text{and} \qquad \overline{\Gamma}_3 = \bigcup_{e_3 \in E_h^{(3)}} \overline{\Gamma}_3^{(e_3)},$$

where $\Gamma_2^{(e_2)}$ and $\Gamma_3^{(e_3)}$ are those edges of the mesh triangles that lie on Γ_2 and Γ_3, respectively. Now we number all edges that lie on Γ_2 and Γ_3 such that we obtain index sets $E_h^{(2)} = \{1, 2, \ldots, N_h^{(2)}\}$ and $E_h^{(3)} = \{1, 2, \ldots, N_h^{(3)}\}$. Furthermore, we define an edge connectivity table $e : i \leftrightarrow \alpha$, with $\alpha \in A^{(e)} = \{1, 2\}$, that relates the local nodal number α to the corresponding global nodal number i of the edge e. For our model problem CHIP, the edge connectivity table (Table 4.3) describes the edges that lie on Γ_3.

Number of edges that lie on Γ_3	Global numbers of the local nodes	
	$P_1^{(e_3)}$	$P_2^{(e_3)}$
1	1	2
2	2	3
3	3	4
4	4	5

Table 4.3. *Connectivity table for the edges that lie on Γ_3.*

The line integrals over Γ_3 (similarly on Γ_2) are calculated as follows:

$$\int_{\Gamma_3} \alpha(x) u_h(x) v_h(x)\, ds = \sum_{e_3 \in E_h^{(3)}} \int_{\Gamma_3^{(e_3)}} \alpha(x) u_h(x) v_h(x)\, ds$$

$$= \sum_{e_3 \in E_h^{(3)}} \sum_{i \in \overline{\omega}_h} \sum_{k \in \overline{\omega}_h} u^{(i)} v^{(k)} \int_{\Gamma_3^{(e_3)}} \alpha(x) \varphi^{(i)}(x) \varphi^{(k)}(x)\, ds.$$

Introducing the index set $\overline{\omega}_h^{(e_3)} = \{j : x^{(j)} = (x_{1,j}, x_{2,j}) \in \overline{\Gamma}_3^{(e_3)}\}$ and taking into account that $\varphi^{(j)}(x)\big|_{\Gamma_3^{(e_3)}} = 0$ for $x^{(j)} \notin \overline{\Gamma}_3^{(e_3)}$, we can rewrite the line integral in the form

$$\int_{\Gamma_3} \alpha(x) u_h(x) v_h(x)\, ds = \sum_{e_3 \in E_h^{(3)}} \sum_{i \in \overline{\omega}_h^{(e_3)}} \sum_{k \in \overline{\omega}_h^{(e_3)}} u^{(i)} v^{(k)} \int_{\Gamma_3^{(e_3)}} \alpha(x) \varphi^{(i)}(x) \varphi^{(k)}(x)\, ds,$$

which, together with the correspondences $e : i \leftrightarrow \alpha$ and $e : k \leftrightarrow \beta$, gives the representation

$$\int_{\Gamma_3} \alpha(x) u_h(x) v_h(x)\, ds = \sum_{e_3 \in E_h^{(3)}} \sum_{\alpha \in A^{(e_3)}} \sum_{\beta \in A^{(e_3)}} u_\alpha^{(e_3)} v_\beta^{(e_3)} \int_{\Gamma_3^{(e_3)}} \alpha(x) \varphi_\alpha^{(e_3)}(x) \varphi_\beta^{(e_3)}(x)\, ds$$

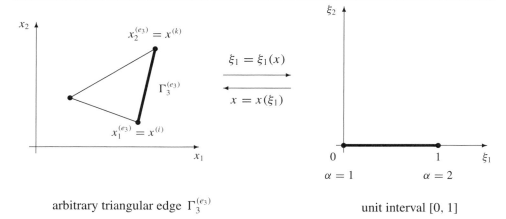

arbitrary triangular edge $\Gamma_3^{(e_3)}$ unit interval $[0, 1]$

Figure 4.14. *Mapping of a triangular edge onto the unit interval* $[0, 1]$.

$$= \sum_{e_3 \in E_h^{(3)}} (v_1^{(e_3)} \; v_2^{(e_3)}) K^{(e_3)} \begin{pmatrix} u_1^{(e_3)} \\ u_2^{(e_3)} \end{pmatrix},$$

with

$$K^{(e_3)} = \left[\int_{\Gamma_3^{(e_3)}} \alpha(x) \varphi_\alpha^{(e_3)}(x) \varphi_\beta^{(e_3)}(x) \, ds \right]_{\beta, \alpha = 1}^{2}.$$

In order to compute the coefficients of the matrix $K^{(e_3)}$, we again transform the line integrals over the edge $\Gamma_3^{(e_3)}$ to line integrals over the unit interval $[0, 1]$ (see Fig. 4.14). This is an affine linear transformation given by the formula

$$x = x(\xi_1) \; : \quad \begin{pmatrix} x_1 \\ x_2 \end{pmatrix} = \begin{pmatrix} x_{1,k} - x_{1,i} \\ x_{2,k} - x_{2,i} \end{pmatrix} \xi_1 + \begin{pmatrix} x_{1,i} \\ x_{2,i} \end{pmatrix}.$$

Using the integral transformation formulas, we obtain

$$\int_{\Gamma_3^{(e_3)}} \alpha(x) \varphi_\alpha^{(e_3)}(x) \varphi_\beta^{(e_3)}(x) \, ds$$

$$= \int_0^1 \alpha(x(\xi_1)) \varphi_\alpha^{(e_3)}(x(\xi_1)) \varphi_\beta^{(e_3)}(x(\xi_1)) \sqrt{\left(\frac{dx_1(\xi_1)}{d\xi_1} \right)^2 + \left(\frac{dx_2(\xi_1)}{d\xi_1} \right)^2} \, d\xi_1$$

$$= \int_0^1 \alpha(x(\xi_1)) \phi_\alpha(\xi_1) \phi_\beta(\xi_1) \sqrt{(x_{1,k} - x_{1,i})^2 + (x_{2,k} - x_{2,i})^2} \, d\xi_1.$$

In the case of linear shape functions, we have

$$\phi_1(\xi_1) = 1 - \xi_1 \qquad \text{and} \qquad \phi_2(\xi_1) = \xi_1.$$

In the same way, we compute the integrals

$$\int_{\Gamma_3} \alpha(x)g_3(x)v_h(x)\,ds \qquad \text{and} \qquad \int_{\Gamma_2} g_2(x)v_h(x)\,ds; \qquad \text{i.e.,}$$

$$\int_{\Gamma_3} \alpha(x)g_3(x)v_h(x)\,ds = \sum_{e_3 \in E_h^{(3)}} \sum_{k \in \overline{\omega}_h^{(e_3)}} v^{(k)} \int_{\Gamma_3^{(e_3)}} \alpha(x)g_3(x)\varphi^{(k)}(x)\,ds$$

$$= \sum_{e_3 \in E_h^{(3)}} \sum_{\beta \in A^{(e_3)}} v_\beta^{(e_3)} \int_{\Gamma_3^{(e_3)}} \alpha(x)g_3(x)\varphi_\beta^{(e_3)}(x)\,ds$$

$$= \sum_{e_3 \in E_h^{(3)}} (v_1^{(e_3)}\ v_2^{(e_3)}) \underline{f}^{(e_3)},$$

with

$$\underline{f}^{(e_3)} = \left[\int_{\Gamma_3^{(e_3)}} \alpha(x)g_3(x)\varphi_\beta^{(e_3)}(x)\,ds \right]_{\beta=1}^2 .$$

Analogously, we obtain

$$\underline{f}^{(e_2)} = \left[\int_{\Gamma_2^{(e_2)}} g_2(x)\varphi_\beta^{(e_2)}(x)\,ds \right]_{\beta=1}^2 .$$

Now we assemble the edge stiffness matrices $K^{(e_3)}$ as well as the edge load vectors $\underline{f}^{(e_3)}$ and $\underline{f}^{(e_2)}$, $e_3 = 1, 2, \ldots, N_h^{(3)}$, $e_2 = 1, 2, \ldots, N_h^{(2)}$, into the global stiffness matrix and into the global load vector in the same way as we assembled $K^{(r)}$ and $\underline{f}^{(r)}$ above.

Including 1st-kind BCs

First Method: Homogenization.
We first put $u^{(j)} = g_1(x^{(j)})$ for all $j \in \gamma_h = \overline{\omega}_h \setminus \omega_h$, i.e., for all indices j belonging to the nodes on the Dirichlet boundary Γ_1, and then we correct the right-hand side as follows:

$$f^{(i)} := f^{(i)} - \sum_{j \in \gamma_h} K_{ij} u^{(j)} \qquad \forall\, i \in \omega_h, \qquad (4.16)$$

where K_{ij} denotes the coefficients of the global stiffness matrix after assembling the element stiffness matrices $K^{(r)}$ and the edge stiffness matrices $K^{(e_3)}$ and $f^{(i)}$ are the coefficients of the global load vector after assembling the element load vectors $\underline{f}^{(r)}$ and the edge load vectors $\underline{f}^{(e_2)}$ and $\underline{f}^{(e_3)}$. Afterwards we cancel the columns j, $j \in \gamma_h$, and the rows i, $i \in \gamma_h$, from the global stiffness matrix and the component with the index i, $i \in \gamma_h$, from the global load vector. The resulting system $K_h \underline{u}_h = \underline{f}_h$ of finite element equations has dimension N_h.

Second Method: Homogenization without row canceling.
Instead of canceling columns and rows in the matrix we apply (4.16) and set, for all $j \in \gamma_h$,

$$f^{(j)} = g_1(x^{(j)}), \quad K_{jj} = 1, \qquad \text{and} \qquad K_{ij} = K_{ji} = 0 \quad \forall\, i \in \omega_h.$$

The resulting system $K_h \underline{u}_h = \underline{f}_h$ of finite element equations has dimension $\bar{N}_h$.

<u>Third Method:</u> Penalty technique (see also [73]).
For all $i \in \gamma_h$, we put

$$K_{ii} := K_{ii} + \hat{K} , \quad \hat{K} = 10^p \cdot \max_{j=1,2,\ldots,\bar{N}_h} |K_{ij}| \cdot \bar{N}_h,$$

$$f^{(i)} := f^{(i)} + \hat{K} g_1(x^{(i)}),$$

with a sufficiently large natural number p. The resulting system $K_h \underline{u}_h = \underline{f}_h$ of finite element equations has dimension $\bar{N}_h$.

4.2.3   Analysis of the Galerkin FEM

The following results can be concluded directly from the Lax–Milgram theory applied to the Galerkin scheme replacing V_0 with V_{0h}. Without loss of generality, we consider again the homogenized *variational problem* (4.6) under our *standard assumptions* (4.7).

Let $V_{0h} = \text{span} \{\varphi^{(i)} : i \in \omega_h\} \subset V_0$ be some finite dimensional subspace of V_0, where $\Phi = \{\varphi^{(i)} : i \in \omega_h\}$ denotes some basis of V_{0h}. More precisely, we think about a family of subspaces with $h \in \Theta$ such that $\dim V_{0h} = N_h \to \infty$ as $h \to 0$ ($|h| \to 0$) (see section 4.2.2 for one example of such finite element spaces).

Theorem 4.5 (Lax–Milgram for the Galerkin Scheme). *Let us assume that our standard assumptions (4.7) are fulfilled and that $V_{0h} \subset V_0$ is a finite dimensional subspace of V_0 as introduced above. Then the following statements hold:*

1. *There exists a unique solution $u_h \in V_{0h}$ of the following Galerkin scheme:*

$$\text{Find } u_h \in V_{0h} : a(u_h, v_h) = \langle F, v_h \rangle \quad \forall v_h \in V_{0h}. \qquad (4.17)$$

2. *The sequence $\{u_h^n\} \subset V_{0h}$, uniquely defined by the following equations:*
   Find $u_h^{n+1} \in V_{0h}$ such that

$$\left(u_h^{n+1}, v_h\right) = \left(u_h^n, v_h\right) - \varrho \left(a \left(u_h^n, v_h\right) - \langle F, v_h \rangle\right) \quad \forall v_h \in V_{0h}, \qquad (4.18)$$

   holds and converges to the solution $u_h \in V_{0h}$ of (4.17) in V_0 ($\| \cdot \|, (\cdot, \cdot)$) for an arbitrary initial guess $u_h^0 \in V_{0h}$ and for some fixed $\varrho \in (0, 2\mu_1/\mu_2^2)$.

3. *The following iteration error estimates are valid:*

   (a) $\|u_h - u_h^n\| \le q^n \|u_h - u_h^0\|$ *(a priori estimate),*

   (b) $\|u_h - u_h^n\| \le \frac{q^n}{1-q} \|u_h^1 - u_h^0\|$ *(a priori bound),*

   (c) $\|u_h - u_h^n\| \le \frac{q}{1-q} \|u_h^n - u_h^{n-1}\|$ *(a posteriori bound),*

   with $0 \le q_{\text{opt}} = q(\varrho_{\text{opt}}) \le q(\varrho) := (1 - 2\mu_1\varrho + \mu_2^2\varrho^2)^{\frac{1}{2}} < 1$ for $\varrho \in (0, 2\mu_1/\mu_2^2)$; $\varrho_{\text{opt}} = \mu_1/\mu_2^2$, $q_{\text{opt}} = \sqrt{1 - \xi^2}$, $\xi = \mu_1/\mu_2$.

The proof of this theorem is analogous to the proof of the Lax–Milgram theorem, Theorem 4.4. Note that $V_{0h} \subset V_0 \subset V$ is also a closed, nontrivial subspace of V. It is easy to see that the iteration process (4.18) is a fixed point iteration that converges for $\varrho \in (0, 2\mu_1/\mu_2^2)$ due to Banach's fixed point theorem [115].

The discretization error estimate in the norm $\| \cdot \|$ of the space V (V_0) is based on Cea's lemma.

Theorem 4.6 (Cea's Lemma). *Let us assume that our standard assumptions (4.7) are fulfilled. Furthermore, we assume that $V_{gh} \subset V_g$ is a finite dimensional submanifold of V_g and that $V_{0h} \subset V_0$ is a finite dimensional subspace of V_0, as introduced above. Then the discretization error estimate*

$$\|u - u_h\| \le \frac{\mu_2}{\mu_1} \inf_{w_h \in V_{gh}} \|u - w_h\| \qquad (4.19)$$

holds, where $u \in V_0$ and $u_h \in V_{0h}$ are the solutions of the abstract variational problem (4.1) and its Galerkin approximation (see section 4.2.1), respectively.

Proof. Consider (4.1) for an arbitrary test function $v_h \in V_{gh} \subset V_g$. If we subtract the Galerkin equations (see scheme given in section 4.2.1) from (4.1), then we obtain the so-called Galerkin orthogonality relation

$$a(u - u_h, v_h) = 0 \quad \forall\, v_h \in V_{0h}. \qquad (4.20)$$

Choosing an arbitrary $w_h \in V_{gh}$ and inserting $v_h = u - u_h - (u - w_h) = w_h - u_h \in V_{0h}$ into (4.20), we get the identity

$$a(u - u_h, u - u_h) = a(u - u_h, u - w_h) \quad \forall\, w_h \in V_{gh}. \qquad (4.21)$$

Now, using the V_0-ellipticity and the V_0-boundedness of the bilinear form $a(\cdot, \cdot)$, we can estimate (4.21) from both above and below as follows:

$$\mu_1 \|u - u_h\|^2 \le a(u - u_h, u - u_h) = a(u - u_h, u - w_h)$$
$$\le \mu_2 \|u - u_h\| \, \|u - w_h\| \quad \forall\, w_h \in V_{gh}.$$

This completes the proof. $\quad\square$

Cea's result means that the discretization error problem with respect to the norm of the space V is reduced to a pure approximation error problem (= stability). Error estimates of the form (4.19) are called quasi optimal because they lead to estimates of discretization error from both above and below:

$$\inf_{w_h \in V_{gh}} \|u - w_h\| \le \|u - u_h\| \le \frac{\mu_2}{\mu_1} \inf_{w_h \in V_{gh}} \|u - w_h\|. \qquad (4.22)$$

It follows immediately from (4.22) that the Galerkin solution u_h converges asymptotically to the solution u of our variational problem (4.1) as the discretization parameter h tends to 0 if and only if $\inf_{w_h \in V_{gh}} \|u - w_h\| \longrightarrow 0$ for $h \longrightarrow 0$. The family $\{V_{0h}\}_{h \in \Theta}$ of finite dimensional subspaces of the space V_0 is called complete in the limit if and only if

$$\lim_{\substack{h \to 0 \\ h \in \Theta}} \inf_{v_h \in V_{0h}} \|u - v_h\| = 0 \quad \forall\, u \in V_0.$$

If $g = 0$, or $g \in V_{gh} \ \forall\, h \in \Theta$, then the limiting completeness ensures the convergence of the Galerkin method.

Let us return to our model problem CHIP described in section 4.2. We obtain a family of triangulation by refining recursively all triangles, e.g., by dividing all triangles into four triangles by connecting the midpoints of the edges. The last procedure results in a family of regular triangulations, ensuring that the edges of the triangles are all of the same order $O(h)$ and the angles of all triangles are not less than some fixed angle $\theta_0 \in (0, \pi/2)$ and not greater than $\pi - \theta_0$. If $u \in H^2(\Omega) \cap V_g$, i.e., u also has quadratically integrable second-order derivatives, then we can prove that there exists an h-independent, positive constant c such that the approximation estimate

$$\inf_{w_h \in V_{gh}} \|u - w_h\|_{1,\Omega} \le c\, h\, \|u\|_{2,\Omega} \tag{4.23}$$

holds, where the norm $\|\cdot\|_{2,\Omega}$ in $H^2(\Omega)$ is defined by

$$\|u\|_{2,\Omega}^2 = \|u\|_{1,\Omega}^2 + \int_\Omega \left[\left(\frac{\partial^2 w}{\partial x_1^2} \right)^2 + 2 \left(\frac{\partial^2 w}{\partial x_1 \partial x_2} \right)^2 + \left(\frac{\partial^2 w}{\partial x_2^2} \right)^2 \right] dx.$$

Cea's estimate (4.19) and the approximation error estimate (4.23) imply the a priori discretization error estimate

$$\|u - u_h\|_{1,\Omega} \le \frac{\mu_2}{\mu_1}\, c\, h\, \|u\|_{2,\Omega}; \tag{4.24}$$

i.e., the convergence rate in the H^1-norm is $O(h)$. However, due to the obtuse angles in the boundary Γ of the computational domain Ω of our model problem CHIP, we can expect less regularity of the solution u than H^2. This reduces the order of convergence to $O(h^\lambda)$ with some positive λ less than 1. Using appropriately graded grids, we can recover the $O(h)$ convergence rate.

Detailed discretization error analysis and discretization error estimates in other norms (e.g., L_2- and L_∞-norms) can be found in [22]. A survey on a posteriori error estimates and adaptive mesh refinement techniques is given in [109].

4.2.4   Iterative solution of the Galerkin system

The convergence rate of classical iterative methods, such as the Richardson method, the Jacobi iteration, and the Gauss–Seidel iteration (cf. Chapter 2), strongly depends on the condition number $\kappa(K_h)$ of the stiffness matrix K_h [80]. In the symmetric positive definite (SPD) case, the spectral condition number is defined by the quotient

$$\kappa(K_h) := \lambda_{max}(K_h) / \lambda_{min}(K_h) \tag{4.25}$$

of the maximal eigenvalue $\lambda_{max}(K_h)$ over the minimal eigenvalue $\lambda_{min}(K_h)$ of the stiffness matrix K_h. The minimal and maximal eigenvalues of the SPD matrix K_h can be characterized by the Rayleigh quotients

$$\lambda_{min}(K_h) := \min_{\underline{v}_h \in \mathbb{R}^{N_h}} \frac{(K_h \underline{v}_h, \underline{v}_h)}{(\underline{v}_h, \underline{v}_h)} \text{ and } \lambda_{max}(K_h) := \max_{\underline{v}_h \in \mathbb{R}^{N_h}} \frac{(K_h \underline{v}_h, \underline{v}_h)}{(\underline{v}_h, \underline{v}_h)}, \tag{4.26}$$

respectively. A careful analysis of the Rayleigh quotients (4.26) leads us to the relations

$$\lambda_{min}(K_h) = O(h^m) \text{ and } \lambda_{max}(K_h) = O(h^{m-2}) \tag{4.27}$$

for the minimal and maximal eigenvalues of the finite element stiffness matrix K_h derived from symmetric, second-order, elliptic BVPs, including our model problem CHIP ($m = 2$), on a regular triangulation, where m denotes the dimension of the computational domain $\Omega \subset \mathbb{R}^m$. This analysis is based on the fundamental relation

$$(K_h \underline{u}_h, \underline{v}_h) = a(u_h, v_h) \ \forall \, \underline{u}_h, \underline{v}_h \in \mathbb{R}^{N_h}, \ \underline{u}_h, \underline{v}_h \longleftrightarrow u_h, v_h \in V_{0h}, \tag{4.28}$$

relating the stiffness matrix K_h to the underlying bilinear form $a(\cdot, \cdot)$, the standard assumption (4.7), and the representation of the integrals over the computational domain Ω as a sum of integrals over all finite elements (triangles). The behavior (4.27) of the minimal and maximal eigenvalues and (4.25) implies that the spectral condition number of the finite element stiffness matrix behaves like

$$\kappa(K_h) = O(h^{-2}) \tag{4.29}$$

as h tends to 0. Therefore, the classical iteration methods mentioned above show a slow convergence on fine grids. More precisely, the number $I(\varepsilon)$ of iterations needed to reduce the initial iteration error by the factor ε^{-1} behaves like $\kappa(K_h) \ln(\varepsilon^{-1})$, i.e., like $h^{-2} \ln(\varepsilon^{-1})$.

To avoid this drawback, say in the case of the classical Richardson method, we need some preconditioning. Let us look at the fixed point iteration process (4.18). Once some basis is chosen, the iteration process (4.18) is equivalent to the iteration method (cf. also (2.8))

$$B_h \frac{\underline{u}_h^{n+1} - \underline{u}_h^n}{\varrho} + K_h \, \underline{u}_h^n = \underline{f}_h, \ n = 0, 1, \ldots, \underline{u}_h^0 \in \mathbb{R}^{N_h}, \tag{4.30}$$

in $\mathbb{R}^{N_h}$, with the SPD matrix (Gram matrix with respect to the scalar product $(\cdot, \cdot)$ in $V_0(V)$)

$$B_h = \left[(\varphi^{(i)}, \varphi^{(k)}) \right]_{k, i \in \omega_h}. \tag{4.31}$$

Indeed, because of the Galerkin–Ritz isomorphism $u_h \leftrightarrow \underline{u}_h$, we have the correspondence

$$u_h^{n+1} \in V_{0h} : (u_h^{n+1}, v_h) = (u_h^n, v_h) - \varrho(a(u_h^n, v_h) - \langle F, v_h \rangle) \quad \forall v_h \in V_{0h}$$

$$u_h^j = \sum_{i \in \omega_h} u_j^{(i)} \varphi^{(i)}, \; j = n, n+1, \text{ where } j \text{ is the iteration index,}$$

$$v_h = \varphi^{(k)}, \; k \in \omega_h,$$

$$\underline{u}_h^{n+1} \in \mathbb{R}^{N_h} :$$

$$\sum_{i \in \omega_h} u_{n+1}^{(i)} \left(\varphi^{(i)}, \varphi^{(k)} \right) = \sum_{i \in \omega_h} u_n^{(i)} \left(\varphi^{(i)}, \varphi^{(k)} \right) - \varrho \sum_{1 \in \omega_h} u_n^{(i)} a \left(\varphi^{(i)}, \varphi^{(k)} \right) + \varrho \left\langle F, \varphi^{(k)} \right\rangle$$

$$\forall k \in \omega_h,$$

$$B_h \underline{u}_h^{n+1} \qquad = B_h \underline{u}_h^n - \varrho K_h \underline{u}_h^n + \varrho \underline{f}_h.$$

Due to the identity

$$\|v_h\| = \|\underline{v}_h\|_{B_h} := (B_h \underline{v}_h, \underline{v}_h)_{\mathbb{R}^{N_h}}^{0.5} \quad \forall v_h \leftrightarrow \underline{v}_h \in \mathbb{R}^{N_h},$$

we can rewrite the iteration error estimates 3(a) to 3(c) of Theorem 4.5 in the B_h-energy norm $\| \cdot \|_{B_h}$, where $(\cdot, \cdot)_{\mathbb{R}^{N_h}}$ denotes the Euclidean inner product.

The iteration (4.30) is *effective* if and only if the system

$$B_h \underline{w}_h^n = \underline{d}_h^n := \underline{f}_h - K_h \underline{u}_h^n,$$
$$\underline{u}_h^{n+1} = \underline{u}_h^n + \varrho \underline{w}_h^n,$$

which has to be handled at each iteration step, can be solved *quickly*; i.e., if we denote by $\text{ops}(B_h^{-1} * \underline{d}_h^n)$ the number of arithmetical operations for solving the system $B_h \underline{w}_h^n = \underline{d}_h^n$, then this effort should be of the order $O(\text{ops}(K_h * \underline{u}_h^n))$ of the number of arithmetical operations needed for one matrix-vector multiplication; e.g., the solver is called asymptotically optimal in this case. Unfortunately, this is *not* true in practical applications. (There exist a few exceptions for special problems, namely, the so-called fast direct methods [80].) Note that the convergence rate $q = q_{\text{opt}} = \sqrt{1 - \xi^2} < 1$ is, independently of h, less than 1, because $\xi = \mu_1/\mu_2$ does not depend on h.

To avoid the drawback of a nonoptimal solver, we introduce an SPD *preconditioner* C_h that should satisfy the following two conditions:

1. $\text{ops}\left(C_h^{-1} * \underline{d}_h^n \right) = O\left(\text{ops}(K_h * \underline{u}_h^n) \right).$
2. C_h is spectrally equivalent to B_h; i.e.,
   $\exists \gamma_1, \gamma_2 = const. > 0 : \gamma_1 C_h \leq B_h \leq \gamma_2 C_h; \text{ i.e.,} \quad \forall \underline{v}_h \in \mathbb{R}^{N_h},$
   $\gamma_1 (C_h \underline{v}_h, \underline{v}_h)_{\mathbb{R}^{N_h}} \leq (B_h \underline{v}_h, \underline{v}_h)_{\mathbb{R}^{N_h}} \leq \gamma_2 (C_h \underline{v}_h, \underline{v}_h)_{\mathbb{R}^{N_h}}; \text{ i.e.,}$
   γ_2/γ_1 should not grow too much for $h \to 0.$ $\qquad (4.32)$

Then we can use the iteration method

$$C_h \frac{\underline{u}_h^{n+1} - \underline{u}_h^n}{\tau} + K_h \underline{u}_h^n = \underline{f}_h, \; n = 0, 1, \ldots, \underline{u}_h^0 \in \mathbb{R}^{N_h}, \qquad (4.33)$$

instead of (4.30). The iteration method (4.33) is sometimes called the preconditioned Richardson iteration. Moreover, under the assumptions (4.32), the following theorem shows that the iteration method (4.33) is *efficient*.

Theorem 4.7 (Preconditioned Richardson Iteration). *Let us assume that our standard assumptions (4.7) are fulfilled and that $V_{0h} \subset V_0$ is a finite dimensional subspace of V_0, as introduced above. Furthermore, the SPD preconditioner C_h should satisfy the assumptions (4.32). Then the following statements hold:*

1. *The iteration (4.33) converges to the solution*

$$\underline{u}_h \in \mathbb{R}^{N_h} : K_h \underline{u}_h = \underline{f}_h,$$
$$\updownarrow$$
$$u_h \in V_{0h} : a(u_h, v_h) = \langle F, v_h \rangle \quad \forall v_h \in V_{0h}$$

for every fixed $\tau \in (0, 2v_1/v_2^2)$, with $v_1 = \mu_1 \gamma_1$ and $v_2 = \mu_2 \gamma_2$.

2. *The following iteration error estimates are valid:*

   a) $\|\underline{u}_h - \underline{u}_h^n\|_{C_h} \le q^n \|\underline{u}_h - \underline{u}_h^0\|_{C_h},$

   b) $\|\underline{u}_h - \underline{u}_h^n\|_{C_h} \le \frac{q^n}{1-q} \|\underline{u}_h^1 - \underline{u}_h^0\|_{C_h},$

   c) $\|\underline{u}_h - \underline{u}_h^n\|_{C_h} \le \frac{q}{1-q} \|\underline{u}_h^n - \underline{u}_h^{n-1}\|_{C_h},$

   with $\|\underline{v}_h\|_{C_h} = (\underline{v}_h, \underline{v}_h)_{C_h}^{0.5} := (C_h \underline{v}_h, \underline{v}_h)_{\mathbb{R}^{N_h}}^{0.5}$

   and $0 \le q_{opt} = q(\tau_{opt}) \le q(\tau) := (1 - 2v_1\tau + v_2^2\tau^2)^{0.5} < 1$
   for $\tau \in (0, 2v_1/v_2^2)$, $\tau_{opt} = v_1/v_2^2$, *and* $q_{opt} = \sqrt{1 - \xi^2}$, $\xi = v_1/v_2$.

3. *If $\gamma_2/\gamma_1 \ne c(h)$, then $I(\epsilon) = O\left(\ln \epsilon^{-1}\right)$ iterations and $Q(\epsilon) = O\left(ops(K_h * \underline{u}_h^n) \cdot \ln \epsilon^{-1}\right)$ arithmetical operations are needed in order to reduce the initial error by some factor $\epsilon \in (0, 1)$; i.e.,*

$$\left\| \underline{u}_h - \underline{u}_h^{I(\epsilon)} \right\|_{C_h} \le \epsilon \left\| \underline{u}_h - \underline{u}_h^0 \right\|_{C_h}.$$

The proof of this theorem is easy. Indeed, replacing $\|v_h\| = \|\underline{v}_h\|_{B_h}$ with $\|\underline{v}_h\|_{C_h}$ and taking into account the inequalities

a) $a(v_h, v_h) = (K_h \underline{v}_h, \underline{v}_h) \ge \mu_1 \|\underline{v}_h\|_{B_h}^2 \ge \mu_1 \gamma_1 \|\underline{v}_h\|_{C_h}^2 \quad \forall \underline{v}_h \in \mathbb{R}^{N_h},$

b) $|a(u_h, v_h)| = |(K_h \underline{u}_h, \underline{v}_h)| \le \mu_2 \|\underline{u}_h\|_{B_h} \|\underline{v}_h\|_{B_h}$
$\le \mu_2 \gamma_2 \|\underline{u}_h\|_{C_h} \|\underline{v}_h\|_{C_h} \quad \forall \underline{u}_h, \underline{v}_h \in \mathbb{R}^{N_h},$

we can easily prove the first two statements of the theorem. Alternatively, we can consider the error iteration scheme

$$\underline{z}_h^{n+1} := \underline{u}_h - \underline{u}_h^{n+1} = \left(I_h - \tau C_h^{-1} K_h\right) \underline{z}_h^n.$$

Then we can directly estimate the iteration error in the C_h-energy norm:

$$\left\| \underline{z}_h^{n+1} \right\|_{C_h}^2 = \left\| \left(I_h - \tau C_h^{-1} K_h\right) \underline{z}_h^n \right\|_{C_h}^2$$
$$= (\underline{z}_h^n, \underline{z}_h^n)_{C_h} - 2\tau (C_h^{-1} K_h \underline{z}_h^n, \underline{z}_h^n)_{C_h} + \tau^2 (C_h^{-1} K_h \underline{z}_h^n, C_h^{-1} K_h \underline{z}_h^n)_{C_h}$$
$$\le (1 - 2\mu_1\gamma_1\tau + (\mu_2\gamma_2)^2\tau^2) \left\| \underline{z}_h^n \right\|_{C_h}^2.$$

The last statement follows from the estimate $q^n \le \epsilon$ if $n \ge \ln \epsilon^{-1} / \ln q^{-1}$.

Remark 4.8.
1. Candidates for C_h are, for instance, the following, practically very important preconditioners:
     - SSOR: see section 2.2.2 and [60, 80, 92];
     - ILU (incomplete LU), MILU (modified incomplete LU factorization), IC (in-
       complete Cholesky), MIC (modified incomplete Cholesky): see section 6.1.2
       and [60, 80, 92];
     - Multigrid preconditioners: see Chapter 7 and [12, 60];
     - Multilevel preconditioners (e.g., BPX (Bramble, Pasciak, and Xu precondi-
       tioner): see [12, 60].

   We will define and consider some of these preconditioning procedures, and their
   parallel versions, in Chapters 6 and 7.

2. It is *not* assumed in Theorem 4.7 that $K_h = K_h^T$. However, we immediately conclude
   from the V_0-ellipticity of $a(\cdot, \cdot)$ that the stiffness matrix K_h is positive definite. In
   practice, the GMRES iteration or other Krylov space methods (see section 6.3.2 and
   [80, 92]) are preferred instead of the Richardson iteration.

3. If the stiffness matrix K_h is SPD, then

     a) the convergence rate estimate can be improved [80, 92]:
        $$\nu_1 C_h \leq K_h \leq \nu_2 C_h \ : \ \tau_{\text{opt}} = 2/(\nu_1 + \nu_2), \ q_{\text{opt}} = \tfrac{1-\xi}{1+\xi}, \ \xi = \tfrac{\nu_1}{\nu_2},$$
     b) the conjugate gradient acceleration is feasible and, of course, used in practice
        (see also section 6.2.1).

   The main topic of the following chapters consists of the parallelization of the iteration
schemes and the preconditioning operations.

Exercises

E4.1. Let us suppose that the solution u of the model problem CHIP (Example 4.3) is
sufficiently smooth, at least in both material subdomains Ω_I and Ω_{II}. Derive the
classical formulation of the problem, i.e., the PDEs in Ω_I and Ω_{II}, the interface
conditions on Γ_{In}, and the BCs on Γ_1, Γ_2, and Γ_3.
Hint: Use (back) integration by parts in the main part.

E4.2. Show that Example 4.3 satisfies the standard assumption formulated in (4.7). Compute
μ_1 and μ_2 such that the quotient μ_2/μ_1 is as small as possible. Show that this example
has a unique solution.
Hint: Use Theorem 4.4.

E4.3. Let us consider the Poisson equation (2.1) in the rectangle $\Omega = (0, 2) \times (0, 1)$ under
homogeneous Dirichlet BCs on $\Gamma_1 := \Gamma$. Derive the variational formulation (4.6) of
this problem and show that the standard assumptions (4.7) are fulfilled.
Hint: You will need Friedrich's inequality
$$\int_\Omega |v|^2 dx \ \leq \ c_F \int_\Omega |\nabla v|^2 dx \ \forall \, v \in H_0^1(\Omega) := \{ v \in V = H^1(\Omega) : v = 0 \, \text{on} \, \Gamma \}$$
to prove the V_0-ellipticity [22].

E4.4. Let us again consider the Poisson equation (2.1) in the rectangle $\Omega = (0, 2) \times (0, 1)$ under homogeneous Dirichlet BCs on $\Gamma_1 := \Gamma$ (see also Exercise E4.3 for the variational formulation). Generate the table of nodal coordinates $i : (x_{1,i}, x_{2,i})$ (cf. Table 4.2) and the table of element connectivities $r : \alpha \leftrightarrow i$ (cf. Table 4.1) for the triangulation derived from Fig. 2.1 by dividing all cells of the mesh into two triangles connecting the bottom-left node with the top-right node of the cell, where the nodes should be numbered from the left to the right and from the bottom to the top. Furthermore, derive the finite element equations $K_h \underline{u}_h = \underline{f}_h$ for linear triangular elements (Courant's elements) according to the element-wise assembling technique presented above. Compare the finite element stiffness matrix and the load vector with their finite difference counterparts given in (2.4).

Hint: Use the homogenization method (first method) for implementing the Dirichlet BCs.

Chapter 5

Basic Numerical Routines in Parallel

*Program testing can be used to show the presence of bugs,
but never to show their absence.*
—Edsger W. Dijkstra (1930–2002)

5.1 Storage of sparse matrices

The finite element method (FEM) introduced in section 4.2 combines the local approximations of the partial differential equation (PDE) into one huge system of linear equations, which is usually extremely *sparse*. The term "sparse" is used for a class of systems of equations, e.g., arising from FEM discretization, where the small number of nonzero entries per row does not grow with the number of equations.

If we think of the matrix as a table and fill all the locations with nonzero elements, then we get the *nonzero pattern of the matrix*. Depending on the properties of that pattern we call a matrix structured or unstructured. The length of the interval from the diagonal element to the last nonzero element of a row/column is called the *band/profile*.

In the following, sparse matrices are always thought of as unstructured; i.e., we will not take advantage of special structures inside sparse matrices.

There exist many storage schemes for matrices and we focus our attention on the *compressed row storage* (*CRS*) scheme. Here only the nonzero elements of a sparse matrix are stored. Therefore, an indirect addressing of the matrix entries is needed. The matrix

$$K_{n \times m} = \begin{pmatrix} 10 & 0 & 0 & -2 \\ 3 & 9 & 0 & 0 \\ 0 & 7 & 8 & 7 \\ 3 & 0 & 8 & 7 \end{pmatrix}$$

can be stored using just two INTEGER vectors and one REAL/DOUBLE vector.

| Values | : | *val* | = |

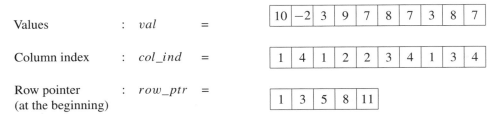

| Column index | : | *col_ind* | = |

| Row pointer (at the beginning) | : | *row_ptr* | = |

The row pointer may also point to the end of the row or, in the case of symmetric matrices, to the diagonal. Further storage schemes for sparse matrices such as, e.g., compressed column storage (CCS), compressed diagonal storage (CDS), jagged diagonal storage (JDS), and skyline storage (SKS), are nicely introduced in [8, §4.3.1]. SKS stores all matrix elements within the variable band/profile of that matrix and usually requires much more memory than CRS. This is most commonly used when direct solvers such as Gaussian elimination and LU factorization are used to solve the system of algebraic equations. The concrete choice of band or profile storage depends on the algorithm used. The proper band/profile will be preserved during the solution process.

More storage schemes for sparse matrices can be found in [8]. A change in the storage scheme will change the routines mentioned in this chapter and can be applied in the following chapters analogously.

5.2   DD by nonoverlapping elements

The data distribution with respect to domain decomposition (DD) uses geometrical connections for the parallelization. The domain Ω is split into P subdomains Ω_i ($i = \overline{1, P}$). Each Ω_i is mapped onto a process. Figure 5.1 presents part of a nonoverlapping DD. We denote edges between subdomains as *interfaces*.

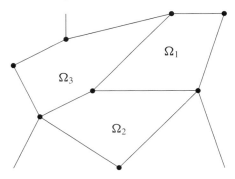

Figure 5.1. *Nonoverlapping DD.*

We distinguish between element-based and node-based data, which we may store in overlapping or nonoverlapping forms (the domains are still nonoverlapped). This leads to four types of data distributions, namely, overlapping elements, overlapping nodes, nonoverlapping elements, and nonoverlapping nodes. We investigate the nonoverlapping distribution of (finite) elements in detail.

Assume that our domain Ω is split into four subdomains, which are discretized by linear triangular finite elements (see Fig. 5.2).

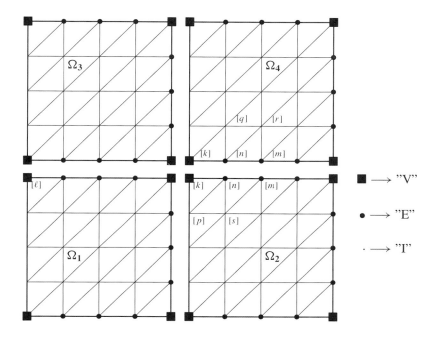

Figure 5.2. *Nonoverlapping elements.*

We distinguish among three sets of nodes denoted by the appropriate subscripts:

"I" nodes located in the interior of a domain $[N_I := \sum_{i=1}^{P} N_{I,i}]$,

"E" nodes located in the interior of an interface edge $[N_E := \sum_{j=1}^{n_e} N_{E,j}]$, and

"V" cross points (vertices), i.e., nodes at start and end of an interface edge $[N_V]$.

The two latter sets are often combined as coupling nodes with the subscript "C" $[N_C = N_V + N_E]$. The total number of nodes is $N = N_V + N_E + N_I = \sum_{i=1}^{P} N_i$.

For simplicity we number the cross points first, then the edge nodes, and finally the inner nodes. The nodes of one edge possess a sequential numbering; the same holds for the inner nodes of each subdomain, so that all vectors and matrices have a block structure of the form

$$\underline{v}^T = (\underline{v}_V, \underline{v}_{E,1}, \ldots, \underline{v}_{E,n_e}, \underline{v}_{I,1}, \ldots, \underline{v}_{I,P})^T.$$

According to the mapping of nodes to the P subdomains $\overline{\Omega}_i$ $(i = 1, 2, \ldots, P)$, all entries of matrices and vectors are distributed on the appropriate process $\mathbb{P}_i$. The coincidence matrices A_i $(i = \overline{1, P})$ represent this mapping and can be considered as a subdomain connectivity matrix, similar to the element connectivities in Table 4.1. In detail, an $N \times N_i$ matrix A_i is a boolean matrix that maps the global vector $\underline{v}$ onto the local vector $\underline{v}_i$.

The following are properties of A:

– Entries for inner nodes appear *exactly once* per row and column.

– Entries for coupling nodes appear once in A_i if this node belongs to process $\mathbb{P}_i$.

Now we define two types of vectors: the accumulated vector (Type I) and the distributed vector (Type II):

Type I: $\underline{u}$ and $\underline{w}$ are stored in process $\mathbb{P}_i (\hat{=} \overline{\Omega}_i)$ such that $\underline{u}_i = A_i \underline{u}$ and $\underline{w}_i = A_i \underline{w}$; i.e., each process $\mathbb{P}_i$ owns the full values of its vector components.

Type II: $\underline{r}, \underline{f}$ are stored as $\underline{r}_i, \underline{f}_i$ in $\mathbb{P}_i$ such that $\underline{r} = \sum_{i=1}^{P} A_i^T \underline{r}_i$ holds; i.e., each node on the interface $(\underline{r}_{C,i})$ owns only its contribution to the full values of its vector components.

Starting with a local FEM discretization of the PDE we naturally achieve a distributed matrix K. If there are convective terms in the PDE and one uses higher-order upwind methods, then additional information is needed from neighboring elements (communication) in order to calculate the local matrix entries.

Matrix K is stored in a distributed way, analogous to a Type II vector, and it is classified as a Type II matrix:

$$K = \sum_{i=1}^{p} A_i^T K_i A_i, \tag{5.1}$$

with K_i denoting the stiffness matrix belonging to subdomain $\overline{\Omega}_i$ (to be more precise: the support of the FEM test functions is restricted to $\overline{\Omega}_i$). If one thinks of $\overline{\Omega}_i$ as one large finite element, then the distributed storing of the matrix is equivalent to the presentation of that element matrix before the fine element accumulation.

The special numbering of nodes implies the following block representation of the equation $K\underline{u} = \underline{f}$:

$$\begin{pmatrix} K_V & K_{VE} & K_{VI} \\ K_{EV} & K_E & K_{EI} \\ K_{IV} & K_{IE} & K_I \end{pmatrix} \begin{pmatrix} \underline{u}_V \\ \underline{u}_E \\ \underline{u}_I \end{pmatrix} = \begin{pmatrix} \underline{f}_V \\ \underline{f}_E \\ \underline{f}_I \end{pmatrix}. \tag{5.2}$$

Therein, K_I is a block diagonal matrix with entries $K_{I,i}$. Similar block structures are valid for $K_{IE}, K_{IE}, K_{IV}, K_{VI}$.

If we really perform the global accumulation of matrix K, then the result is a Type I matrix $\mathfrak{M}$ and we can write

$$\mathfrak{M}_i := A_i \mathfrak{M} A_i^T. \tag{5.3}$$

Although $K = \mathfrak{M}$ holds, we need to distinguish between both representations because of the different local storing (i.e., $K_i \neq \mathfrak{M}_i$).

The diagonal matrix

$$R = \sum_{i=1}^{p} A_i^T A_i \qquad (5.4)$$

contains for each node the number of subdomains it belongs to (priority of a node); e.g., in Fig. 5.2, $R^{[k]} := R_{kk} = 4$, $R^{[n]} = R^{[m]} = R^{[p]} = 2$, $R^{[q]} = 1$.

We use subscripts and superscripts in the remaining section in the following way: $v_{C,i}^{[n]}$ denotes the nth component (in local or global numbering) of vector $\underline{r}$ stored in process $\mathbb{P}_i$. Subscript "C" indicates a subvector belonging to the interface. A similar notation is used for matrices.

5.3 Vector-vector operations

Change of vector types

Obviously, we can apply addition/subtraction and similar operations without communication on vectors of the same type.

The conversion of a Type II (distributed) vector into a Type I (accumulated) vector requires communication:

$$\underline{\mathfrak{w}}_i = A_i \sum_{i=1}^{p} A_i^T \underline{r}_i. \qquad (5.5)$$

The other direction, conversion of a Type I vector into a Type II vector, is not unique. One possibility consists of locally dividing each vector component by its priority (number of subdomains a node belongs to); i.e.,

$$\underline{r}_i = R^{-1} \underline{\mathfrak{w}}_i, \qquad (5.6)$$

with R defined in (5.4).

Inner product

The inner product of two vectors of *different type* requires communication only for the global sum of one scalar:

$$(\underline{\mathfrak{w}}, \underline{r}) = \underline{\mathfrak{w}}^T \underline{r} = \underline{\mathfrak{w}}^T \sum_{i=1}^{p} A_i^T \underline{r}_i = \sum_{i=1}^{p} (A_i \underline{\mathfrak{w}})^T \underline{r}_i = \sum_{i=1}^{p} (\underline{\mathfrak{w}}_i, \underline{r}_i). \qquad (5.7)$$

Every other combination of vectors requires a conversion of the type, possibly with communication.

5.4   Matrix-vector operations

Here, we investigate matrix-vector multiplication with matrices and vectors of various types. A more abstract presentation of this topic with proofs of the following statements can be found in [48, 52, 56].

1. *Type* II *matrix* × *Type* I *vector results in a Type* II *vector.*
   Indeed, we achieve, from definition (5.1),

$$
\mathsf{K} \cdot \underline{\mathfrak{w}} \;=\; \sum_{i=1}^{p} A_i^T \mathsf{K}_i A_i \cdot \underline{\mathfrak{w}} \;=\; \sum_{i=1}^{p} A_i^T \underbrace{\mathsf{K}_i \cdot \underline{\mathfrak{w}}_i}_{\underline{r}_i} \;=\; \underline{r}. \tag{5.8}
$$

   The (not necessary) execution of the summation (implying communication) results in a Type I vector.

2. *Type* II *matrix* × *Type* II *vector* needs a conversion of the vector before the multiplication in item 1.

3. The operation *Type* I *matrix* × *Type* I *vector cannot be performed with arbitrary Type* I *matrices* $\mathfrak{M}$.
   For a detailed analysis, we study the operation $\underline{u} = \mathfrak{M} \cdot \underline{\mathfrak{w}}$ on node n in Fig. 5.2. The local multiplication $\underline{u}_i = \mathfrak{M}_i \cdot \underline{\mathfrak{w}}_i$ should result in a Type I vector $\underline{u}$ at node n in both subdomains:

$$
u_2^{[n]} = \mathfrak{M}_2^{[n,n]} \mathfrak{w}_2^{[n]} + \mathfrak{M}_2^{[n,m]} \mathfrak{w}_2^{[m]} + \mathfrak{M}_2^{[n,k]} \mathfrak{w}_2^{[k]} + \mathfrak{M}_2^{[n,p]} \mathfrak{w}_2^{[p]} + \mathfrak{M}_2^{[n,s]} \mathfrak{w}_2^{[s]},
$$

$$
u_4^{[n]} = \mathfrak{M}_4^{[n,n]} \mathfrak{w}_4^{[n]} + \mathfrak{M}_4^{[n,m]} \mathfrak{w}_4^{[m]} + \mathfrak{M}_4^{[n,k]} \mathfrak{w}_4^{[k]} + \mathfrak{M}_4^{[n,q]} \mathfrak{w}_4^{[q]} + \mathfrak{M}_4^{[n,r]} \mathfrak{w}_4^{[r]}.
$$

   Terms 4 and 5 of the right-hand side in the above equations differ such that processes 2 and 4 achieve different results instead of a unique one. The reason lies in the transport of information via the matrix from a process $\mathbb{P}_a$ to a node belonging additionally to another process $\mathbb{P}_b$.

   Denote by $\sigma^{[i]} = \{s \,:\, x_i \in \overline{\Omega}_s\}$ all processors a node i belongs to; e.g., in Fig. 5.2, $\sigma^{[n]} = \{2, 4\}$, $\sigma^{[p]} = \{1, 2\}$, $\sigma^{[s]} = \{2\}$. Then the transport of information from node j to a node i is only admissible if $\sigma^{[i]} \subseteq \sigma^{[j]}$. Representing the matrix entries as a directed graph, then, e.g., the entries $q \rightarrow k$, $q \rightarrow n$, $s \rightarrow n$, $s \rightarrow m$, $s \rightarrow p$, $p \rightarrow k$, $n \rightarrow k$, $p \rightarrow n$, $n \rightarrow p$, $\ell \rightarrow k$, $k \rightarrow \ell$ in Fig. 5.2 are not admissible.

   As shown in [52], the operation $\mathfrak{M} \cdot \underline{u}$ can be performed without communication only for the following special structure:

$$
\mathfrak{M} = \begin{pmatrix} \mathfrak{M}_V & 0 & 0 \\ \mathfrak{M}_{EV} & \mathfrak{M}_E & 0 \\ \mathfrak{M}_{IV} & \mathfrak{M}_{IE} & \mathfrak{M}_I \end{pmatrix} \quad \Longrightarrow \quad \underline{u} = \mathfrak{M} \cdot \underline{\mathfrak{w}}, \tag{5.9}
$$

   with block diagonal matrices $\mathfrak{M}_I$, $\mathfrak{M}_E$, and $\mathfrak{M}_V$. In particular, the submatrices must not contain entries between nodes belonging to different sets of processors. The proof for (5.9) can be found in [53, 54].

The block diagonality of $\mathfrak{M}_I$ and the correct block structure of $\mathfrak{M}_{IE}$ and $\mathfrak{M}_{IV}$ are guaranteed by the data decomposition. However, the mesh must fulfill the following two requirements in order to guarantee the correct block structure for the remaining three matrices:

a) There is no connection between vertices belonging to different sets of subdomains. This guarantees the block diagonality of $\mathfrak{M}_V$ (often $\mathfrak{M}_V$ is a diagonal matrix). There is no connection between vertices and opposite edges; e.g., in Fig. 5.1, $\ell \nrightarrow p$. This ensures the admissible structure of $\mathfrak{M}_{EV}$ (and $\mathfrak{M}_{VE}$).

b) There is no connection between edge nodes belonging to different sets of subdomains. This guarantees the block diagonality of $\mathfrak{M}_E$.

Requirement (a) can easily be fulfilled if at least one node is located on all edges between two vertices and one node lies in the interior of each subdomain (or has been added therein). A modification of a given mesh generator also guarantees that requirement (b) is fulfilled; e.g., the edge between nodes p and n is omitted. Figure 5.3 presents the revised mesh.

- This strategy can be applied only to triangular or tetrahedral finite elements.

- In general, rectangular elements do not fulfill condition (b).

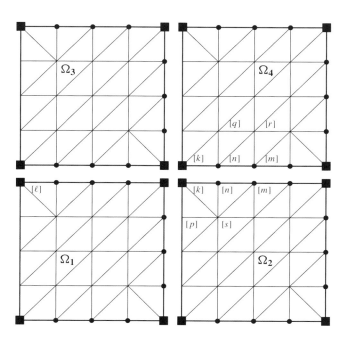

Figure 5.3. *Nonoverlapping elements with a revised discretization.*

4. Similarly to the previous point, *Type* I *matrix* × *Type* II *vector cannot be performed with general Type* I *matrices* $\mathfrak{M}$. But we would expect a Type II vector as a result.

Assuming that the necessary requirements (a) and (b) are fulfilled by the mesh (see Fig. 5.3), we focus our interest on the operation $\underline{f}_I = M_{IC} \cdot \underline{r}_C$. Performing this operation locally on processor 4 gives

$$\underline{f}_4^{[r]} = \mathfrak{M}_{IC,4}^{[r,n]} \cdot \underline{r}_4^{[n]} + \mathfrak{M}_{IC,4}^{[r,m]} \cdot \underline{r}_4^{[m]}.$$

Due to the Type II vector property, we are missing the entries $\mathfrak{M}_{IC,2}^{[r,n]} \cdot \underline{r}_2^{[n]}$ and $\mathfrak{M}_{IC,2}^{[r,m]} \cdot \underline{r}_2^{[m]}$ in the result. The difference with respect to the previous point is that the transport of information from node j to a node i is permitted only if $\sigma^{[i]} \supseteq \sigma^{[j]}$. Thus, only a matrix of the shape

$$\mathfrak{M} = \begin{pmatrix} \mathfrak{M}_V & \mathfrak{M}_{VE} & \mathfrak{M}_{VI} \\ 0 & \mathfrak{M}_E & \mathfrak{M}_{EI} \\ 0 & 0 & \mathfrak{M}_I \end{pmatrix} \implies \underline{f} = \sum_{i=1}^{P} A_i^T \underline{f}_i = \sum_{i=1}^{P} A_i^T \left(\mathfrak{M}_i \underline{r}_i \right),$$
(5.10)

with block diagonal matrices $\mathfrak{M}_I, \mathfrak{M}_E, \mathfrak{M}_V$, can be used in this type of matrix-vector multiplication.

Remark 5.1. Items 3 and 4 can be combined under the assumption that $\mathfrak{M}_V, \mathfrak{M}_E, \mathfrak{M}_I$ are block diagonal matrices (guaranteed by the mesh in Fig. 5.3). Denote by $\mathfrak{M}_L, \mathfrak{M}_U$, and $\mathfrak{M}_D$ the strictly lower, upper, and diagonal parts of $\mathfrak{M}$, respectively. Then we can perform the Type I matrix-vector multiplication for all types of vectors:

$$\underline{w} = \mathfrak{M} \cdot \underline{u} := (\mathfrak{M}_L + \mathfrak{M}_D) \cdot \underline{u} + \sum_{i=1}^{P} A_i^T \mathfrak{M}_{U,i} R_i^{-1} \cdot \underline{u}_i,$$
(5.11a)

$$\underline{w} = \mathfrak{M} \cdot \underline{r} := (\mathfrak{M}_L + \mathfrak{M}_D) \sum_{i=1}^{P} A_i^T \cdot \underline{r}_i + \sum_{i=1}^{P} A_i^T \mathfrak{M}_{U,i} \cdot \underline{r}_i,$$
(5.11b)

$$\underline{f} = \mathfrak{M} \cdot \underline{u} := R^{-1}(\mathfrak{M}_L + \mathfrak{M}_D) \cdot \underline{u} + \mathfrak{M}_U R^{-1} \cdot \underline{u},$$
(5.11c)

$$\underline{f} = \mathfrak{M} \cdot \underline{r} := R^{-1}(\mathfrak{M}_L + \mathfrak{M}_D) \sum_{i=1}^{P} A_i^T \cdot \underline{r}_i + \mathfrak{M}_U \cdot \underline{r}.$$
(5.11d)

Note that in any case the above multiplications require two type conversions, but the choice of the vectors directly influences the amount of communication needed.

If the matrix $\mathfrak{M}$ is available as a factor product $\mathfrak{M} = \mathfrak{L}^{-1}\mathfrak{U}^{-1}$, with the lower and the upper triangular matrices $\mathfrak{L}^{-1}$ and $\mathfrak{U}^{-1}$ fulfilling conditions a) and b), then we write

$$\underline{w} = \mathfrak{L}^{-1}\mathfrak{U}^{-1} \cdot \underline{r} := \mathfrak{L}^{-1} \sum_{i=1}^{P} A_i^T \mathfrak{U}_i^{-1} \cdot \underline{r}_i,$$
(5.12a)

$$\underline{f} = \mathfrak{L}^{-1}\mathfrak{U}^{-1} \cdot \underline{u} := R^{-1}\mathfrak{L}^{-1} \sum_{i=1}^{P} A_i^T \mathfrak{U}_i^{-1} R^{-1} \cdot \underline{u}_i.$$
(5.12b)

The second case, i.e., $M = \mathfrak{U}^{-1}\mathfrak{L}^{-1}$, results in

$$\underline{\mathfrak{w}} = \mathfrak{U}^{-1}\mathfrak{L}^{-1} \cdot \underline{r} := \sum_{i=1}^{P} A_i^T \mathfrak{U}_i^{-1} R_i^{-1} A_i \mathfrak{L}^{-1} \cdot \left(\sum_{j=1}^{P} A_j^T \underline{r}_j \right), \qquad (5.13a)$$

$$\underline{f} = \mathfrak{U}^{-1}\mathfrak{L}^{-1} \cdot \underline{u} := \mathfrak{U}^{-1} R^{-1} \mathfrak{L}^{-1} \cdot \underline{u}. \qquad (5.13b)$$

The equations for the remaining combinations of vector types in (5.12) and (5.13) can easily be derived by type conversion. Here the number and the kind of type conversion are also determined by the factoring method used with the matrix $\mathfrak{M}$.

   A similar methodology can be applied if other data distributions, e.g., overlapping nodes, are considered.

Exercises

Let some double precision vector $\underline{u}$ be stored block-wise disjoint, i.e., distributed over all processes s ($s = 0, \ldots, P - 1$) such that $\underline{u} = (\underline{u}_0^T, \ldots, \underline{u}_{P-1}^T)^T$.

E5.1. Write a routine

DebugD(nin, xin, icomm)

that prints *nin* double precision numbers of the array *xin*. Start the program with several processes. As a result, all processes will write their local vectors at roughly the same time. Hence, you must look carefully through the output for the data of process s.

   Improve the routine **DebugD** so that process 0 reads the number (from terminal) of the process that should write its vector. Use

MPI_Bcast

to broadcast this information and let the processes react appropriately. If necessary use **MPI_Barrier** to synchronize the output.

E5.2. Exchange the global minimum and maximum of the vector $\underline{u}$. Use

MPI_Gather, MPI_Scatter/ MPI_Bcast, and ExchangeD.

How can you reduce the amount of communication?

Hint: Compute the local minimum/maximum. Afterwards let some process determine the global quantities.

   Alternatively, you can use **MPI_Allreduce** and the operation **MPI_Minloc** or **MPI_Maxloc**.

E5.3. Write a routine

Skalar(n_s, x, y, icomm)

to compute the global scalar product $\langle x, y \rangle$ of two double precision vectors x and y with local lengths n_s. Use

MPI_Allreduce with the operation MPI_SUM.

procx*(procy−1)		procx*procy−1
procx		
0	1	procx−1

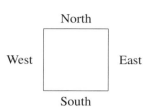

Let the unit square $[0, 1]^2$ be partitioned uniformly into $procx \times procy$ rectangles Ω_i numbered row by row. The numbering of the subdomains coincides with the corresponding process IDs (or ranks in MPI jargon).

The function

IniGeom(myid, procx, procy, neigh, color)

from *example/accuc* generates the topological relations corresponding to the DD defined above. This information is stored in the integer array *neigh*(4). A checkerboard coloring is defined in *color*. Moreover, the function

IniCoord(myid, procx, procy, xl, xr, yb, yt)

can be used to generate the coordinates of the lower-left corner (xl, yb) and the upper-right corner (xr, yt) of each subdomain.

E5.4. Implement a local data exchange of a double precision number between each processor and all of its neighbors (connected by a common edge). Use the routine **ExchangeD** from **E3.7**.

Let each subdomain Ω_i be uniformly discretized into $nx * ny$ rectangles generating a triangular mesh (nx, ny are the same for all subdomains); see Fig. 5.4.

If we use linear finite element test functions, then each vertex of the triangles represents one component of the solution vector, e.g., the temperature at this point, and we have $nd := (nx + 1) * (ny + 1)$ local unknowns within one subdomain. We propose a local ordering in rows of the unknowns. Note that the global number of unknowns is $N = (procx * nx + 1) * (procy * ny + 1) < (procx * procy * nd)$.

The function

GetBound(id, nx, ny, w, s)

copies the values of w corresponding to the boundaries south (id = 1), east (id = 2), north (id = 3), west (id = 4) into the auxiliary vector s. Conversely, the function

AddBound(id, nx, ny, w, s)

adds the values of s to the components of w corresponding to the nodes on the boundaries south (id = 1), east (id = 2), north (id = 3), and west (id = 4). These functions can be used for the accumulation (summation) of values corresponding to the nodes on the interfaces between two adjacent subdomains, which is a typical and necessary operation.

E5.5. Write a routine that accumulates a distributed double precision vector w. The call to such a routine could look like

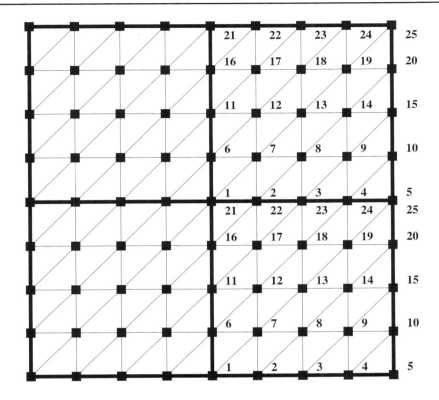

Figure 5.4. *Four subdomains in local numbering with local discretization $nx =$ $ny = 4$ and global discretization $N_x = N_y = 8$.*

VecAccu(nx, ny, w, neigh, color, myid, icomm)

where w is both the input and the output vector.

E5.6. Use the routine from **E5.5** to determine the number of subdomains sharing a node.

E5.7. Write a routine that implements the parallel multiplication of a distributed matrix in CRS format with an accumulated vector.

E5.8. Determine the accumulated diagonal of a distributed matrix in CRS format.

Chapter 6

Classical Solvers

The Analytical Engine has no pretensions whatever to originate anything. It can do whatever we know how to order it to perform.
—Ada Augusta, Countess of Lovelace (1815–1851)

In this chapter we solve $n \times n$ sparse systems of linear equations

$$K\underline{u} = \underline{f}. \tag{6.1}$$

We explicitly assume that the system of equations above was generated by discretizing an elliptic partial differential equation (PDE). Matrix K is sparse if the finite element method (FEM), finite difference method (FDM), or finite volume method (FVM) has been used for discretization. In the following, we assume exactly this sparsity pattern of the matrix. A detailed description of numerical algorithms used in this chapter can be found in [2, 8, 80]. A different introduction to parallelization of these algorithms is given in [26].

All parallel algorithms refer to the *nonoverlapping element distribution* and basic routines in sections 5.2 to 5.4. The appropriate vector and matrix types are denoted in the same way as therein. The use of any other data distribution in the parallelization will be emphasized explicitly.

If we know in advance the amount of arithmetic work needed to solve (6.1), not taking into account round-off errors, then we call this solution method a direct one. Otherwise, it is an iterative method, since the number of iterations is data dependent.

Gaussian elimination could be used in principle to solve (6.1), which results from a discretization with a typical discretization parameter h in the $m = 2, 3$ dimensional space. Unfortunately the solution time is of order $O(n^3) = O(h^{-3m})$ for dense matrices; i.e., doubling the number of unknowns increases the solution time by a factor of 8. Even if we take into account the bandwidth of sparse matrices as $BW = O(h^{-m+1})$, then the arithmetical costs are only reduced to $O(n \cdot BW^2) = O(h^{-3m+2})$. Therefore, some iterative techniques are introduced that require for sparse systems only $O(h^{-m-2} \ln(\varepsilon))$ operations (with an achieved accuracy of ε at termination). It is even possible to achieve $O(n) = O(h^{-m})$ operations (see Chapter 7).

6.1 Direct methods

6.1.1 LU factorization

Let K be a dense matrix, row-wise stored, L a lower triangular matrix with normalization $\ell_{i,i} = 1$, column-wise stored, and U an upper triangular matrix, row-wise stored, and let $\sum_{j=1}^{n} L_{i,j} \cdot U_{j,k} = K_{i,k}$. We investigate the rank-r modification in Algorithm 6.1 for the LU factorization without pivot search, as depicted in Fig. 6.1.

```
DO i = 1, n
    DO k = i, n                              Determine ith row of U.
        U_{i,k} := K_{i,k}
    END DO
    L_{i,i} := 1
    DO k = i + 1, n                          Determine ith column of L.
        L_{k,i} := K_{k,i}/U_{i,i}
    END DO
    DO k = i + 1, n                          Transform rest of matrix.
        DO j = i + 1, n
            K_{k,j} := K_{k,j} - L_{k,i} · U_{i,j}
        END DO
    END DO
END DO
```

Algorithm 6.1: Rank-r modification of LU factorization—sequential.

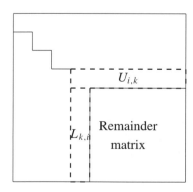

Figure 6.1. *Illustration of the rank-r modification.*

Parallelization of the LU factorization

For the purpose of parallelization on a distributed memory computer, a block variant of the rank-r modification, as in Algorithm 6.2, seems preferable. Now $K_{k,i}$ denotes the rows and columns of the block splitting of matrix K.

DO $i = 1, n$
    a) Factor $K_{k,i}$ such that $K_{k,i} = L_{k,i} \cdot U_{i,i}$ $(k = \overline{i, n})$
      $\longrightarrow U_{i,i}, L_{i,i}, L_{i+1,i}, \ldots, L_{n,i}$.
    b) Determine $U_{i,\ell}$ such that $K_{i,\ell} = L_{i,i} \cdot U_{i,\ell}$ $(\ell = \overline{i+1, n})$
      $\longrightarrow U_{i,i+1}, \ldots, U_{i,n}$.
    c) Rank-r modification of remainder matrix
      $K_{k,\ell} := K_{k,\ell} - L_{k,i} \cdot U_{i,\ell}$ $(k, \ell = \overline{i+1, n})$.
END DO

Algorithm 6.2: Block variant of rank-r modification.

If we use a row-wise or column-wise *distribution of matrix K* on the processors, then the smaller the remainder matrix gets, the fewer processors are used (see Fig. 6.1). Therefore, we use a square block scattered decomposition as it is implemented in a parallel version of ScaLAPACK (see Fig. 6.2).

$K_{1,1}$ $\mathbb{P}_0$	$K_{1,2}$ $\mathbb{P}_1$	$K_{1,3}$ $\mathbb{P}_2$	$K_{1,4}$ $\mathbb{P}_0$	$K_{1,5}$ $\mathbb{P}_1$
$K_{2,1}$ $\mathbb{P}_3$	$K_{2,2}$ $\mathbb{P}_4$	$K_{2,3}$ $\mathbb{P}_5$	$K_{2,4}$ $\mathbb{P}_3$	$K_{2,5}$ $\mathbb{P}_4$
$K_{3,1}$ $\mathbb{P}_6$	$K_{3,2}$ $\mathbb{P}_7$	$K_{3,3}$ $\mathbb{P}_8$	$K_{3,4}$ $\mathbb{P}_6$	$K_{3,5}$ $\mathbb{P}_7$
$K_{4,1}$ $\mathbb{P}_0$	$K_{4,2}$ $\mathbb{P}_1$	$K_{4,3}$ $\mathbb{P}_2$	$K_{4,4}$ $\mathbb{P}_0$	$K_{4,5}$ $\mathbb{P}_1$
$K_{5,1}$ $\mathbb{P}_3$	$K_{5,2}$ $\mathbb{P}_4$	$K_{5,3}$ $\mathbb{P}_5$	$K_{5,4}$ $\mathbb{P}_3$	$K_{5,5}$ $\mathbb{P}_4$

5×5 matrix blocks

3×3 Processors $(i = \overline{0.n})$

$K_{i,j}$
$\downarrow$
$\mathbb{P}_{(i-1) \mod P_x, (j-1) \mod P_y}$

Figure 6.2. *Scattered distribution of a matrix.*

Now we can formulate the *parallel block version* by using the scattered distribution of K, see Algorithm 6.3.

Nonblocking communication seems advantageous from the viewpoint of implementation. A distribution of blocks in the direction of rows and columns similar to the hypercube numbering (ring is embedded in hypercube [75, 89]) is also possible.

6.1.2 ILU factorization

If we use the LU factorization from section 6.1.1 for solving (6.1), then the matrix K will be factored into an upper triangular matrix $\widetilde{U}$ and a lower triangular matrix $\widetilde{L}$. Solving the resulting graduated system of equations with the triangular matrices is fast, but the time for

Process $\mathbb{P}(K_{i,j})$ stores block $K_{i,j}$.

DO $i = 1, n$
a) $\mathbb{P}(K_{i,i})$:
        Determine $L_{i,i}$ and $U_{i,i}$ from $K_{i,i} = L_{i,i} \cdot U_{i,i}$.
        Send $L_{i,i}$ to all $\mathbb{P}(K_{i,\ell})$ via the ring in direction of columns ($\ell = \overline{i+1, n}$).
        Send $U_{i,i}$ to all $\mathbb{P}(K_{k,i})$ via the ring in direction of rows ($k = \overline{i+1, n}$).
b) $\mathbb{P}(K_{k,i})$ [$k = \overline{i+1, n}$]:
        Determine $L_{k,i}$ from $K_{k,i} = L_{k,i} \cdot U_{i,i}$.
        Send $L_{k,i}$ to all $\mathbb{P}(K_{k,\ell})$ via the ring in direction of columns ($\ell = \overline{i+1, n}$).
    $\mathbb{P}(K_{i,\ell})$ [$\ell = \overline{i+1, n}$]
        Determine $U_{i,\ell}$ from $K_{i,\ell} = L_{i,i} \cdot U_{i,\ell}$.
        Send $U_{i,\ell}$ to all $\mathbb{P}(K_{k,\ell})$ via the ring in direction of rows ($k = \overline{i+1, n}$).
c) $\mathbb{P}(K_{k,\ell})$ [$k, \ell = \overline{i+1, n}$]
        $K_{k,\ell} := K_{k,\ell} - L_{k,i} \cdot U_{i,\ell}$
END DO

Algorithm 6.3: Block-wise rank-r modification of LU factorization — parallel.

the factorization

$$K =: \widetilde{L}\widetilde{U} \tag{6.2}$$

behaves like $O(n \cdot BW^2)$. If the matrix is sparse (FEM, FDM, . . .), then the fill-in in the factorization increases the memory requirements unacceptably when storing the factored matrix.

Therefore, one tries to factor matrix K into triangular matrices L and U with the same matrix pattern as K. Equation (6.2) has to be modified by taking the remainder matrix P into account:

$$K = LU - P. \tag{6.3}$$

There exist special modifications for symmetric positive definite (SPD) matrices (IC = Incomplete Cholesky) and for factorizations allowing a certain fill-in in the pattern of the triangular matrices (ILU(m)) [26, 105]. They are often used in combination with lumping techniques; i.e., entries not fitting in the admissible pattern are added to the main diagonal (e.g., MAF [81]).

We investigate in the following the incomplete factorizations (6.3) for nonsymmetric matrices K without fill-in, i.e., the classical ILU(0) factorization, and the preconditioning step $LU\underline{w} = \underline{r}$.

Sequential algorithm

We write the block-wise LU factorization of a matrix K using the block structure of vectors and matrices introduced in (5.2) from section 5.2. The remainder matrices P in Algorithm 6.4 are used only for notational purposes and they are omitted in the implementation. Pointwise ILU is used within the single blocks. After factorization in Algorithm 6.4, the solution is derived using Algorithm 6.5.

$$
\begin{array}{ll}
\text{Determine} & \\
L_V, U_V & K_V = L_V \cdot U_V - P_V \\
L_{EV} & K_{EV} = L_{EV} \cdot U_V - P_{EV} \\
L_{IV} & K_{IV} = L_{IV} \cdot U_V - P_{IV} \\
U_{VE} & K_{VE} = L_V \cdot U_{VE} - P_{VE} \\
U_{VI} & K_{VI} = L_V \cdot U_{VI} - P_{VI} \\
\hline
 & K_E := K_E - L_{EV} \cdot U_{VE} \\
 & K_{EI} := K_{EI} - L_{EV} \cdot U_{VI} \\
 & K_{IE} := K_{IE} - L_{IV} \cdot U_{VE} \\
 & K_I := K_I - L_{IV} \cdot U_{VI} \\
\hline
\text{Determine} & \\
L_E, U_E & K_E = L_E \cdot U_E - P_E \\
L_{IE} & K_{IE} = L_{IE} \cdot U_E - P_{IE} \\
U_{EI} & K_{EI} = L_E \cdot U_{EI} - P_{EI} \\
\hline
 & K_I := K_I - L_{IE} \cdot U_{EI} \\
\hline
\text{Determine} & \\
L_I, U_I & K_I = L_I \cdot U_I - P_I
\end{array}
$$

Algorithm 6.4: Sequential block ILU factorization.

$$
\begin{array}{lll}
\text{I)} & \text{Solve } L\underline{u} = \underline{r} & \\
 & \underline{u}_V := & L_V^{-1}\underline{r}_V \\
 & \underline{u}_E := & L_E^{-1}(\underline{r}_E - L_{EV}\underline{u}_V) \\
 & \underline{u}_I := & L_I^{-1}(\underline{r}_I - L_{IE}\underline{u}_E - L_{IV}\underline{u}_V) \\
\text{II)} & \text{Solve } U\underline{w} = \underline{u} & \\
 & \underline{w}_I := & U_I^{-1}\underline{u}_I \\
 & \underline{w}_E := & U_E^{-1}(\underline{u}_E - U_{EI}\underline{w}_I) \\
 & \underline{w}_V := & U_V^{-1}(\underline{u}_V - U_{VE}\underline{w}_E - U_{VI}\underline{w}_I)
\end{array}
$$

Algorithm 6.5: Sequential block-wise substitution steps for $LU\underline{w} = \underline{r}$.

Parallel ILU Factorization

The first idea for parallelization of ILU is the simple rewriting of Algorithm 6.4 with an accumulated matrix $\mathfrak{K}$ (see Algorithm 6.6). Naturally, $\mathfrak{L}$ and $\mathfrak{U}$ are also accumulated matrices. However, due to (5.9) and (5.10), we have to fulfill requirements 3(a) and 3(b) (from section 5.4) on the mesh. The redundantly stored matrices $\mathfrak{K}_V$, $\mathfrak{K}_E$, $\mathfrak{K}_E$, $\mathfrak{K}_{EV}$, and $\mathfrak{K}_{VE}$ are not updated before their (pointwise) factorization so that the matrix accumulation can be performed in the beginning.

The forward and backward substitution step $\underline{\mathfrak{w}} = \mathfrak{U}^{-1} \cdot \mathfrak{L}^{-1} \cdot \underline{r}$ contains the matrix-vector multiplications (5.9) and (5.10) requiring three total vector type changes. The diagonal matrix R, which includes the number of subdomains sharing a node, was defined in (5.4).

The disadvantage of Algorithm 6.7 is that the wrong vector type is always available. This leads to three vector type changes including two accumulations and means doubling the

Start	$\mathfrak{K} = \sum\limits_{i=1}^{P} A_i^T \mathsf{K}_i A_i$	
Determine		Why parallel ?
$\mathfrak{L}_V, \mathfrak{U}_V$	$\mathfrak{K}_V = \mathfrak{L}_V \cdot \mathfrak{U}_V - P_V$	mesh $\to$ diagonal matrix
$\mathfrak{L}_{EV}$	$\mathfrak{K}_{EV} = \mathfrak{L}_{EV} \cdot \mathfrak{U}_V - P_{EV}$	domain decomposition (DD),
		mesh $\to$ block matrices
L_{IV}	$K_{IV} = L_{IV} \cdot \mathfrak{U}_V - P_{IV}$	DD, mesh $\to$ block matrices
$\mathfrak{U}_{VE}$	$\mathfrak{K}_{VE} = \mathfrak{L}_V \cdot \mathfrak{U}_{VE} - P_{VE}$	DD, mesh $\to$ block matrices
U_{VI}	$\mathfrak{K}_{VI} = \mathfrak{L}_V \cdot U_{VI} - P_{VI}$	DD, mesh $\to$ block matrices
Modify		
	$\mathfrak{K}_E := \mathfrak{K}_E - \mathfrak{L}_{EV} \cdot \mathfrak{U}_{VE}$	same matrix type
	$K_{EI} := K_{EI} - \mathfrak{L}_{EV} \cdot U_{VI}$	same matrix type
	$K_{IE} := K_{IE} - L_{IV} \cdot \mathfrak{U}_{VE}$	same matrix type
	$K_I := K_I - L_{IV} \cdot U_{VI}$	same matrix type
Determine		
$\mathfrak{L}_E, \mathfrak{U}_E$	$\mathfrak{K}_E = \mathfrak{L}_E \cdot \mathfrak{U}_E - P_E$	DD, mesh $\to$ block matrices
L_{IE}	$K_{IE} = L_{IE} \cdot \mathfrak{U}_E - P_{IE}$	DD $\to$ block matrices
U_{EI}	$K_{EI} = \mathfrak{L}_E \cdot U_{EI} - P_{EI}$	DD $\to$ block matrices
Modify		
	$K_I := K_I - L_{IE} \cdot U_{EI}$	same matrix type
Determine		
$\mathfrak{L}_I, \mathfrak{U}_I$	$\mathfrak{K}_I = \mathfrak{L}_I \cdot \mathfrak{U}_I - P_I$	DD $\to$ block matrices

Algorithm 6.6: Parallelized $\mathfrak{L}\mathfrak{U}$ factorization.

I)	$\underline{r} := \sum\limits_{i=1}^{P} A_i^T \underline{r}_i$		IV)	$\underline{w}_I := U_I^{-1}\underline{u}_I$
II)	$\underline{u}_V := \mathfrak{L}_V^{-1}\underline{r}_V$			$\underline{w}_E := \mathfrak{U}_E^{-1}(\underline{u}_E - U_{EI}\underline{w}_I)$
	$\underline{u}_E := \mathfrak{L}_E^{-1}(\underline{r}_E - \mathfrak{L}_{EV}\underline{u}_V)$			$\underline{w}_V := \mathfrak{U}_V^{-1}(\underline{u}_V - \mathfrak{U}_{VE}\underline{w}_E$
	$\underline{u}_I := L_I^{-1}(\underline{r}_I - L_{IE}\underline{u}_E - L_{IV}\underline{u}_V)$			$\phantom{\underline{w}_V := \mathfrak{U}_V^{-1}(\underline{u}_V} - U_{VI}\underline{w}_I)$
III)	$\underline{u} := R^{-1}\underline{u}$		V)	$\underline{w} := \sum\limits_{i=1}^{P} A_i^T \underline{w}_i$

Algorithm 6.7: Parallel forward and backward substitution step for $\mathfrak{L}\mathfrak{U}\underline{w} = \underline{r}$.

communication cost per iteration compared to the ω-Jacobi iteration. This communication behavior can be improved. According to (5.12a), the substitution step $\underline{w} = \mathfrak{L}^{-1} \cdot \mathfrak{U}^{-1} \cdot \underline{r}$ should require only one vector type change (including one accumulation, as demonstrated in the next paragraph).

Parallel IUL factorization

In order to reduce the communication in Algorithm 6.7 we use an incomplete $\mathfrak{U}\mathfrak{L}$ factorization, where the unknowns are passed in the reverse order. Thus, the redundant stored matrices $\mathfrak{K}_V$, $\mathfrak{K}_E$, $\mathfrak{K}_E$, $\mathfrak{K}_{EV}$, and $\mathfrak{K}_{VE}$ are updated locally during the factorization in

Algorithm 6.8, so that their accumulation is only allowed before their factorization. The solution step in Algorithm 6.9 is a simple application of (5.12a).

Start	K	
Determine		Why parallel ?
U_I, L_I	$K_I = U_I \cdot L_I - P_I$	DD $\rightarrow$ block matrices
L_{IE}	$K_{IE} = U_I \cdot L_{IE} - P_{IE}$	DD $\rightarrow$ block matrices
L_{IV}	$K_{IV} = U_I \cdot L_{IV} - P_{IV}$	DD $\rightarrow$ block matrices
U_{EI}	$K_{EI} = U_{EI} \cdot L_I - P_{EI}$	DD $\rightarrow$ block matrices
U_{VI}	$K_{VI} = U_{VI} \cdot L_I - P_{VI}$	DD $\rightarrow$ block matrices
Modify		
	$K_E := K_E - U_{EI} \cdot L_{IE}$	same matrix type
	$K_{EV} := K_{EV} - U_{EI} \cdot L_{IV}$	same matrix type
	$K_{VE} := K_{VE} - U_{VI} \cdot L_{IE}$	same matrix type
	$K_V := K_V - U_{VI} \cdot L_{IV}$	same matrix type
Accumulate $\mathfrak{K}_E, \mathfrak{K}_{EV}, \mathfrak{K}_{VE}$		
e.g.,	$\mathfrak{K}_{EV} := \sum_{i=1}^{P} A_{E,i}^T K_{EV,i} A_{V,i}$	—
Determine		
$\mathfrak{U}_E, \mathfrak{L}_E$	$\mathfrak{K}_E = \mathfrak{U}_E \cdot \mathfrak{L}_E - P_E$	DD, mesh $\rightarrow$ block matrices
$\mathfrak{U}_{VE}$	$\mathfrak{K}_{VE} = \mathfrak{U}_{VE} \cdot \mathfrak{L}_E - P_{VE}$	DD, mesh $\rightarrow$ block matrices
$\mathfrak{L}_{EV}$	$\mathfrak{K}_{EV} = \mathfrak{U}_E \cdot \mathfrak{L}_{EV} - P_{EV}$	DD, mesh $\rightarrow$ block matrices
Modify		
	$K_V := K_V - \mathfrak{U}_{VE} \cdot R_E^{-1} \cdot \mathfrak{L}_{EV}$	same matrix type (5.13a)
Accumulate $\mathfrak{K}_V$		—
Determine		
$\mathfrak{U}_V, \mathfrak{L}_V$	$\mathfrak{K}_V = \mathfrak{U}_V \cdot \mathfrak{L}_V - P_V$	mesh $\rightarrow$ diagonal matrix

Algorithm 6.8: Parallel IUL factorization.

I) $\underline{u}_I := U_I^{-1} \underline{r}_I$	III) $\underline{w}_V := \mathfrak{L}_V^{-1} \underline{u}_V$
$\underline{u}_E := \mathfrak{U}_E^{-1}(\underline{r}_E - U_{EI}\underline{u}_I)$	$\underline{w}_E := \mathfrak{L}_E^{-1}(\underline{u}_E - \mathfrak{L}_{EV}\underline{w}_V)$
$\underline{u}_V := \mathfrak{U}_V^{-1}(\underline{r}_V - \mathfrak{U}_{VE}\underline{u}_E - U_{VI}\underline{u}_I)$	$\underline{w}_I := L_I^{-1}(\underline{u}_I - L_{IE}\underline{w}_E - L_{IV}\underline{w}_V)$
II) $\underline{u} := \sum_{i=1}^{P} A_i^T \underline{u}_i$	

Algorithm 6.9: Parallel backward and forward substitution step for $\mathfrak{U}\,\mathfrak{L}\,\underline{w} = \underline{r}$.

Comparison of ILU and IUL factorization

The two algorithms introduced in the previous two sections differ mainly in the communication behavior sampled in Table 6.1.

The accumulation of the matrix is split into separate calls for accumulation of cross points and interface nodes. Therefore, the communication loads in Algorithms 6.8 and 6.6

	ILU: $\mathfrak{L}\,\mathfrak{U}$	IUL: $\mathfrak{U}\,\mathfrak{L}$
start up	matrix accumulation	—
factorization	—	matrix accumulation
substitution	$2 \times$ vector accumulation	$1 \times$ vector accumulation

Table 6.1. *Communication in ILU and IUL.*

are the same. According to the finite element discretization, the stiffness matrix K is stored distributed (section 5.2), so that we cannot save any communication.

With respect to the communication, we recommend using IUL factorization in parallel implementations.

IUL factorization with reduced pattern

The requirements on the matrix from section 5.4 cannot be fulfilled in the 3D (three-dimensional) case or the 2D case with bilinear/quadratic/... elements. Instead of the original matrix $\mathfrak{K}$ we have to factor a nearby matrix $\widetilde{\mathfrak{K}}$ with the proper reduced pattern. This matrix $\widetilde{\mathfrak{K}}$ can be derived by deleting all entries not included in the reduced pattern or by lumping these entries. We also have to change the internal storage pattern of $\widetilde{\mathfrak{K}}$ (usually stored in some index arrays) in any case, since only setting the nonadmissible entries to zero causes errors in the factorization of $\widetilde{\mathfrak{K}}$.

6.2   Smoothers

In the multigrid literature, the term "smoother" has become synonymous with all iterative methods. However, it is a term that has been abused frequently by many authors (all of us included at one time or another). It was used in [14] to describe the effect of one iteration of an iterative method on each of the components of the error vector.

For many relaxation methods, the norm of each error component is reduced in each iteration: hence the term "smoother". This means that, if the error components are each thought of as wave functions, then the waves are smoothed out somewhat at each iteration. The most effective smoothers are variants of Gaussian elimination, which are one-iteration smoothers (note that Gaussian elimination is not usually referred to as a smoother in the multigrid literature, but as a direct method instead). Other examples of smoothers include Jacobi, Gauss–Seidel, underrelaxation, and alternating direction implicit (ADI).

Not all iterative methods are smoothers, however. Some are roughers [28, 29]. For many iterative methods, while the norm of the error vector is reduced at each iteration, the norm of some of the components of the error may grow at each iteration: hence the term "rougher". For example, all conjugate gradient like methods, symmetric successive overrelaxation (SSOR), and overrelaxation methods are roughers. Roughers are treated in section 6.3.

6.2.1 ω-Jacobi iteration

Sequential algorithm

Let us denote the main diagonal of K by $D = \text{diag}(K)$ and the residual and correction by $\underline{r}$ and $\underline{w}$, respectively. Then we can write the ω-Jacobi iteration using the $KD^{-1}K$-norm in the stopping criterion in the following way:

$$
\begin{aligned}
&\text{Choose } \underline{u}^0 \\
&\underline{r} \;:=\; \underline{f} - K \cdot \underline{u}^0 \\
&\underline{w} \;:=\; D^{-1} \cdot \underline{r} \\
&\sigma \;:=\; \sigma_0 \;:=\; (\underline{w}, \underline{r}) \\
&k \;:=\; 0 \\
&\text{while} \quad \sigma > \text{tolerance}^2 \cdot \sigma_0 \quad \text{do} \\
&\qquad k \;:=\; k + 1 \\
&\qquad \underline{u}^k \;:=\; \underline{u}^{k-1} + \omega \cdot \underline{w} \\
&\qquad \underline{r} \;:=\; \underline{f} - K \cdot \underline{u}^k \\
&\qquad \underline{w} \;:=\; D^{-1} \cdot \underline{r} \\
&\qquad \sigma \;:=\; (\underline{w}, \underline{r}) \\
&\text{end}
\end{aligned}
$$

Algorithm 6.10: Sequential Jacobi iteration.

Data flow of Jacobi iteration

There are no data dependencies between any components of vector $\underline{u}^k$ (kth iterative solution), so that the parallelization of that algorithm is quite easy. The data flow chart is presented in Fig. 6.3.

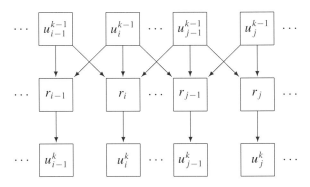

Figure 6.3. *Data flow of Jacobi iteration.*

Parallel algorithm

The parallelization strategy is based on nonoverlapping elements.

- Store matrix K as distributed matrix K (5.1). Otherwise, we have to assume restrictions on the matrix pattern.

- Vectors $\underline{u}^k \rightarrow \underline{u}^k$ are accumulated (Type I) stored. Therefore, the matrix-vector product results in a Type II vector without any communication (5.8). Thus, vectors $\underline{r}, \underline{f} \rightarrow \underline{r}, \underline{f}$ are distributed (Type II) stored and are never accumulated during the iteration.

- By using an accumulated vector $\underline{w}$ we can perform the update from $\underline{u}^{k-1} \rightarrow \underline{u}^k$ by means of vector operations without any communication.

- Due to the different vector types used in the inner product $(\underline{w}, \underline{r})$, that operation requires only a very small amount of communication (5.7).

- Calculation of the correction $\underline{w}$ needs the inversion of D. This can only be done on an accumulated matrix. The calculation of

$$\mathfrak{D}^{-1} = \mathrm{diag}^{-1}\left(\sum_{s=1}^{P} A_s^T \mathrm{K}_s A_s\right) = \left(\sum_{s=1}^{P} A_s^T \mathrm{diag}\{\mathrm{K}_s\} A_s\right)^{-1}$$

requires one communication step in the iteration setup.

- $\mathfrak{D}^{-1}$ is a diagonal matrix and therefore we can apply $\mathfrak{D}^{-1}$ not only to an accumulated vector but also to a distributed one. This leads to the following (numerically and in terms of cost) equivalent formulations for the parallel calculation of the correction:

$$\underline{w} := \sum_{s=1}^{P} A_s^T \mathfrak{D}_s^{-1} \cdot \underline{r}_s = \mathfrak{D}^{-1} \cdot \sum_{s=1}^{P} A_s^T \cdot \underline{r}_s,$$

requiring only one communication per iteration sweep.

$$\boxed{\begin{array}{ll}
\mathfrak{D}^{-1} := \mathrm{diag}^{-1}\left(\sum_{s=1}^{P} A_s^T \mathrm{K}_s A_s\right) \\
\text{Choose } \underline{u}^0 \\
\underline{r} \quad := \underline{f} - \mathrm{K} \cdot \underline{u}^0 \\
\underline{w} \quad := \mathfrak{D}^{-1} \cdot \sum_{s=1}^{P} A_s^T \underline{r}_s \\
\sigma \quad := \sigma_0 := (\underline{w}, \underline{r}) \\
k \quad := 0 \\
\text{while} \quad \sigma > \text{tolerance}^2 \cdot \sigma_0 \quad \text{do} \\
\qquad k \quad := k+1 \\
\qquad \underline{u}^k := \underline{u}^{k-1} + \omega \cdot \underline{w} \\
\qquad \underline{r} \quad := \underline{f} - \mathrm{K} \cdot \underline{u}^k \\
\qquad \underline{w} \quad := \mathfrak{D}^{-1} \cdot \sum_{s=1}^{P} A_s^T \underline{r}_s \\
\qquad \sigma \quad := (\underline{w}, \underline{r}) \\
\text{end}
\end{array}}$$

Algorithm 6.11: Parallel Jacobi iteration: Jacobi($K, \underline{u}^0, \underline{f}$).

If we store the diagonal matrix $\mathfrak{D}^{-1}$ as a vector $\underline{\mathfrak{d}}$, then the operation $\mathfrak{D}^{-1} \cdot (\sum_{s=1}^{P} A_s^T \underline{r}_s)$ has to be implemented as a vector accumulation (communication) followed by the component-wise multiplication of the two accumulated vectors and results in a vector of the same type. Each step of the Jacobi iteration requires only one vector accumulation and one ALL_REDUCE of a real number in the inner product.

If we define $\underline{\mathfrak{w}} := \sum_{s=1}^{P} A_s^T \underline{r}_s$ as an auxiliary vector, it changes the update step into

$$\underline{u}^k := \underline{u}^{k-1} + \omega \cdot D^{-1} \cdot \underline{\mathfrak{w}} = \underline{u}^{k-1} + \omega \cdot \underline{\mathfrak{d}} \circledast \underline{\mathfrak{w}}.$$

Here we denote by $\underline{\mathfrak{d}} \circledast \underline{\mathfrak{w}}$ the component-wise multiplication of two vectors, i.e., $\{\mathfrak{d}_i \cdot \mathfrak{w}_i\}_{i=\overline{1,n}}$, which represents a multiplication of a vector by a diagonal matrix.

The calculation of the inner product is no longer necessary if the Jacobi iteration is used with a fixed number of iterations (this is typical for multigrid smoothers). This saves the ALL_REDUCE operation in the parallel code, and vectors $\underline{r}$ and $\underline{\mathfrak{w}}$ can be stored in one place.

6.2.2 Gauss–Seidel iteration

Denoting by E and F the strict lower and upper triangular submatrices and by D the main diagonal of the sparse matrix K from (6.1) we can write $K = E + D + F$. We use a residual norm in the stopping criterion.

Sequential algorithm

$$
\begin{array}{ll}
\text{Choose } \underline{u}^0 \\
\underline{r} \; := \; \underline{f} - K \cdot \underline{u}^0 \\
\sigma \; := \; \sigma_0 := (\underline{r}, \underline{r}) \\
k \; := \; 0 \\
\text{while} \quad \sigma > \text{tolerance}^2 \cdot \sigma_0 \quad \text{do} \\
\qquad k \; := \; k + 1 \\
\qquad \underline{u}^k \; := \; \underline{u}^{k-1} + D^{-1} \cdot \left(\underline{f} - E \cdot \underline{u}^k - (D+F) \cdot \underline{u}^{k-1} \right) \\
\qquad \underline{\widetilde{r}} \; := \; \underline{f} - K \cdot \underline{u}^k \\
\qquad \sigma \; := \; (\underline{\widetilde{r}}, \underline{\widetilde{r}}) \\
\text{end}
\end{array}
$$

Algorithm 6.12: Sequential Gauss–Seidel forward iteration.

The reason for the generally better convergence behavior of the Gauss–Seidel iteration compared to the ω-Jacobi iteration (Algorithm 6.10) consists of the continuous update of the residual whenever a new component $\underline{u}_i^k$ of the kth iterate has been calculated:

$$r_i := f_i - \sum_{j=1}^{i-1} K_{i,j} u_j^k - \sum_{j=i}^{n} K_{i,j} u_j^{k-1},$$

$$u_i^k := u_i^{k-1} + K_{i,i}^{-1} \cdot r_i. \tag{6.4}$$

Now the pointwise residual r_i and therefore the iterate u_i^k in (6.4) depends on the numbering of the unknowns in vector $\underline{u}$. Hence, the whole Gauss–Seidel iteration depends on the numbering of the unknowns.

In order to save arithmetic work, the K^2-norm $(\widetilde{r}, \widetilde{r})$ is replaced with the norm using the pointwise residual, namely, $(\underline{r}, \underline{r})$.

Data flow of the Gauss–Seidel forward iteration

As presented in Algorithm 6.12 and Fig. 6.4 the components of $\underline{u}^k$ are coupled; e.g., u_i^k cannot be calculated before u_{i-1}^k is determined. In general, all u_s^k $(s < i)$ for which $K_{i,s} \neq 0$ (graph/pattern of matrix) have to be calculated before u_i^k can be updated. Due to the coupling of the components of $\underline{u}^k$ the direct parallelization becomes fine grain and therefore the pointwise Gauss–Seidel iteration is nearly impossible to parallelize.

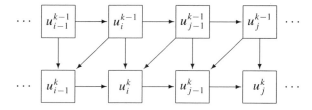

Figure 6.4. *Data flow of Gauss–Seidel forward iteration.*

Red-black Gauss–Seidel iteration

To get rid of the unfavorable properties with respect to the data flow we have to split the subscripts of the vector into at least two adjoint subsets ω_{red} and ω_{black}, so that $K_{i,j} \equiv 0 \; \forall \, i \neq j \in \omega_{\text{red}}$ (resp., ω_{black}). This results in no coupling between elements of the subsets. This allows us to reformulate the update step in Algorithm 6.12 into Algorithm 6.13. The proper data flow is presented in Fig. 6.5.

$$
\begin{aligned}
\underline{u}_{\text{red}}^k &:= \underline{u}_{\text{red}}^{k-1} + D_{\text{red}}^{-1} \cdot \left(\underline{f}_{\text{red}} - (E_{rb} + F_{rb}) \cdot \underline{u}_{\text{black}}^{k-1} - D_{\text{red}} \cdot \underline{u}_{\text{red}}^{k-1} \right) \\
&= D_{\text{red}}^{-1} \cdot \left(\underline{f}_{\text{red}} - (E_{rb} + F_{rb}) \cdot \underline{u}_{\text{black}}^{k-1} \right) \\
\underline{u}_{\text{black}}^k &:= \underline{u}_{\text{black}}^{k-1} + D_{\text{black}}^{-1} \cdot \left(\underline{f}_{\text{black}} - (E_{br} + F_{br}) \cdot \underline{u}_{\text{red}}^k - D_{\text{black}} \cdot \underline{u}_{\text{black}}^{k-1} \right) \\
&= D_{\text{black}}^{-1} \cdot \left(\underline{f}_{\text{black}} - (E_{br} + F_{br}) \cdot \underline{u}_{\text{red}}^k \right)
\end{aligned}
$$

Algorithm 6.13: Update step in red-black Gauss–Seidel forward iteration.

Due to the decoupling property within the subsets of "red" and "black" data, Algorithm 6.13 can be parallelized. This pointwise red/black splitting can be applied to a five-point stencil discretization in two dimensions and also to a seven-point stencil discretization in three dimensions.

A block version of the iteration above is recommended in the parallel case. The blocks should be chosen so that the update of blocks of one color requires no communication.

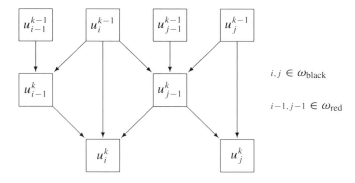

Figure 6.5. *Data flow of red-black Gauss–Seidel forward iteration.*

This produces a checkerboard coloring of the unit square coinciding with the adjoint data distribution (again, two colors are sufficient when a five-point stencil is used).

Parallel algorithms

The update step of the Gauss–Seidel iteration is the significant difference with respect to the Jacobi iteration, so we restrict further investigations to it. A data distribution with nonoverlapping elements (section 5.2) is assumed and represented in Fig. 6.6.

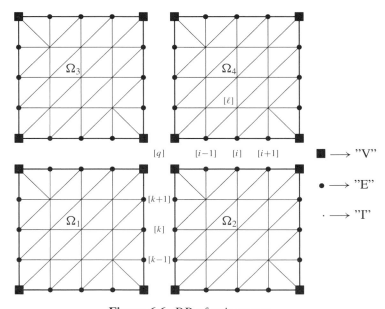

Figure 6.6. *DD of unit square.*

- A formal application of the parallelization strategy used for the ω-Jacobi iteration results in a distributed matrix K, an accumulated diagonal matrix $\mathfrak{D}$, distributed stored vectors $\underline{f}$, and $\underline{r}$, and an accumulated vector $\underline{u}^k$.

- The update step requires vector accumulation.

- The inner product requires different vector types.

- Therefore, we choose an accumulated auxiliary vector $\underline{w} = \sum_{s=1}^{P} A_s^T \underline{r}_s$ fulfilling the two previous requirements.

- The derived algorithm changes only slightly if we define $\underline{w}$ as a correction (as was done in section 6.2.1).

The parallelization of the original Gauss–Seidel method has to solve the problem of data dependencies. Moreover, the processors would have to wait for each other depending on the global numbering of the data (crucial dependency). This effect can be reduced by a special block-wise ordering (coloring) of the nodes.

We use the marked nodes in Fig. 6.6 to illustrate these data dependencies and algorithmical modifications. First we order the nodes (and the unknowns connected with them) such that the vertex nodes (V') are followed by the edge-wise ordered nodes (E) and then by the interior nodes (I); i.e., $q < \cdots < k-1 < k < k+1 < \cdots < i-1 < i < i+1 < \cdots < \ell$.

1. If there are no matrix connections between $k+1$ and $i-1$ (valid for linear test functions on the triangles), then we have no crucial data dependencies. However, we need the accumulated value $\underline{u}_i^k$ (and so on) for the calculation of $\underline{u}_{i+1}^k$. This results in a fine grain parallel Gauss–Seidel iteration on each edge.

2. If we regard the nodes $i-1, i, i+1$ and $k-1, k, k+1$ in each case as one block, then we can formulate a parallel block Gauss–Seidel iteration. There is no matrix connection admissible between both blocks in the example.

3. To overcome the above restriction on the matrix pattern, we combine a block Gauss–Seidel iteration with Jacobi iterations on the interface nodes V and E. This version will now be presented.

Variant: Gauss–Seidel ω-Jacobi

Between the blocks V, E, I and inside the subdomains (block I) a Gauss–Seidel iteration is performed. Inside the blocks V, E a Jacobi iteration is used. Due to the partial use of the Jacobi iteration, the convergence rate is in general worse than in the usual Gauss–Seidel iteration. Operation $\underline{\mathfrak{d}} \circledast \underline{w}$ in Algorithm 6.14 denotes component-wise multiplication of two vectors. We also took into account that in the interior of the domains (I) accumulated and distributed vectors/matrices are identical. The accumulation of cross points (V) and interface data (E) is usually performed separately, so that we need exactly the same amount of communication as in the ω-Jacobi iteration (Algorithm 6.11).

$$\underline{d} := \operatorname{diag}(K^{[i,i]})_{i=\overline{1,n}}$$

$$\underline{\mathfrak{d}} := \sum_{s=1}^{P} A_s^T \underline{d}_s$$

$$\underline{\mathfrak{d}} := \{\omega/d^{[i]}\}_{i=\overline{1,n}}$$

Choose $\underline{u}^0$

$$\underline{r} := \underline{f} - \mathsf{K} \cdot \underline{u}^0$$

$$\underline{\mathfrak{w}} := \sum_{s=1}^{P} A_s^T \underline{r}_{-s}$$

$$\sigma := \sigma_0 := (\underline{\mathfrak{w}}, \underline{r})$$

$$k := 0$$

while $\quad \sigma > \text{tolerance}^2 \cdot \sigma_0 \quad$ do

$\qquad k \quad := k + 1$

$\qquad \underline{r}_V := \underline{f}_V - \mathsf{K}_V \cdot \underline{u}_V^{k-1} - \mathsf{K}_{VE} \cdot \underline{u}_E^{k-1} - \mathsf{K}_{VI} \cdot \underline{u}_I^{k-1}$

$\qquad \underline{\mathfrak{w}}_V := \sum_{s=1}^{P} A_{V,s}^T \underline{r}_{-V,s}$

$\qquad \underline{u}_V^k := \underline{u}_V^{k-1} + \underline{\mathfrak{d}}_V \circledast \underline{\mathfrak{w}}_V$

$\qquad \underline{r}_E := \underline{f}_E - \mathsf{K}_{EV} \cdot \underline{u}_V^k - \mathsf{K}_E \cdot \underline{u}_E^{k-1} - \mathsf{K}_{EI} \cdot \underline{u}_I^{k-1}$

$\qquad \underline{\mathfrak{w}}_E := \sum_{s=1}^{P} A_{E,s}^T \underline{r}_{E,s}$

$\qquad \underline{u}_E^k := \underline{u}_E^{k-1} + \underline{\mathfrak{d}}_E \circledast \underline{\mathfrak{w}}_E$

$\qquad \underline{r}_I := \underline{f}_I - K_{IV} \cdot \underline{u}_V^k - K_{IE} \cdot \underline{u}_E^k$

$\qquad u_{I,i}^k := u_{I,i}^{k-1} + d_{I,i} \cdot \left(r_{I,i} - \sum_{j=N_E+1}^{i-1} K_{I,ij} \cdot u_{I,j}^k - \sum_{j=i}^{N} K_{I,ij} \cdot u_{I,j}^{k-1} \right)$

$\qquad \underline{w}_I := \underline{r}_I - K_I \cdot \underline{u}_I^k$

$\qquad \sigma := (\underline{\mathfrak{w}}, \underline{r})$

end

Algorithm 6.14: Parallel Gauss–Seidel ω-Jacobi forward iteration.

6.2.3 ADI methods

ADI methods are useful when a regular mesh is employed in two or more space dimensions. The first paper that ever appeared with ADI in it was [32] rather than the commonly referenced [85]. The general ADI results for N space variables are found in [31]. The only paper ever written that shows convergence without resorting to requiring commutativity of operators was written by Pearcy [86]. A collection of papers were written applying ADI to FEMs. An extensive treatment can be found in [30]. Wachspress [110] has a very nice treatise on ADI.

The algorithm is dimension dependent, though the basic techniques are similar. The sparse matrix K is decomposed into a sum of matrices that can be permuted into tridiagonal form. These tridiagonal systems can be solved easily, e.g., by the Prongonka algorithm [94]. The permutation normally requires a data transpose that is painful to perform on a parallel computer with distributed memory.

In this section, we investigate ADI for 2D problems on a square domain with a uniform or tensor product mesh. The techniques used work equally well with general line relaxation iterative methods.

Sequential algorithm

Let us begin with the 2D model problem (2.1). Suppose that we have $K = H + V$, where H and V correspond to the discretization in the horizontal and vertical directions only. Hence, H corresponds to the discretization of the term $\frac{\partial^2 u(x,y)}{\partial x^2}$ in (2.2) and V corresponds to the discretization of the term $\frac{\partial^2 u(x,y)}{\partial y^2}$ in (2.3).

$$
\begin{aligned}
&\text{Choose } \underline{u}^0 \\
&\underline{r} \quad := \quad \underline{f} - K \cdot \underline{u}^0 \\
&\sigma \quad := \quad \sigma_0 := (\underline{r}, \underline{r}) \\
&k \quad := \quad 0 \\
&\text{while} \quad \sigma > \text{tolerance}^2 \cdot \sigma_0 \quad \text{do} \\
&\qquad (H + \rho_{k+1}I)\underline{u}^{k*} \quad = \quad (\rho_{k+1}I - V) \cdot \underline{u}^k + f \\
&\qquad (V + \rho_{k+1}I)\underline{u}^{k+1} \quad = \quad (\rho_{k+1}I - H) \cdot \underline{u}^{k*} + f \\
&\qquad \underline{r} \qquad\qquad := \quad \underline{f} - K \cdot \underline{u}^{k+1} \\
&\qquad \sigma \qquad\qquad := \quad (\underline{r}, \underline{r}) \\
&\qquad k \qquad\qquad := \quad k + 1 \\
&\text{end}
\end{aligned}
$$

Algorithm 6.15: Sequential ADI in two dimensions.

The parameters ρ_k are acceleration parameters that are chosen to speed up the convergence rate of ADI. For (2.1), we know the eigenvalues and eigenvectors for K:

$$K\mu_{jm} = \lambda_{jm}\mu_{jm}, \quad \text{where } \mu_{jm} = \sin(j\pi y)\sin(m\pi x/2).$$

We also know that

$$H\mu_{jm} = \lambda_m\mu_{jm} \quad \text{and} \quad V\mu_{jm} = \lambda_j\mu_{jm},$$

where $\lambda_j + \lambda_m = \lambda_{jm}$.

It is also readily verified that H and V commute: $HV = VH$. This condition can be met in practice for all separable elliptic equations with an appropriate discretization of the problem. The error iteration matrix

$$T_\rho = (V + \rho I)^{-1}(H - \rho I)(H + \rho I)^{-1}(V - \rho I)$$

satisfies

$$T_\rho\mu_{jm} = \frac{(\lambda_j - \rho)(\lambda_m - \rho)}{(\lambda_j + \rho)(\lambda_m + \rho)}\mu_{jm}.$$

Suppose that

$$0 < \alpha \le \lambda_j, \lambda_m \le \beta.$$

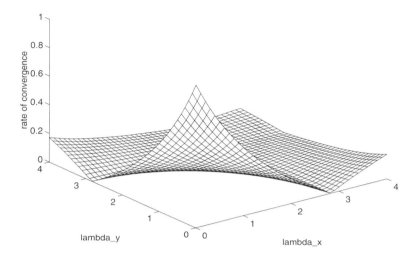

Figure 6.7. *ADI damping factor parameter space.*

A plot (see Fig. 6.7) of which eigenvalues provide good or bad damping of the errors leads us to note that at the extreme values of the eigenvalues there is not much damping, but there is an impressive amount at some point in between.

For a sequence of length γ of acceleration parameters, we define

$$c = \alpha/\beta, \quad \delta = (\sqrt{2} - 1)^2, \quad \text{and } n = \lceil \log_\delta c \rceil + 1.$$

Cyclically choose

$$\rho_j = \beta c^{\frac{j-1}{n-1}}, \quad j = 1, \ldots, n. \tag{6.5}$$

Then the error every n iterations is reduced by a factor of δ. For (2.1) with an $(N_x + 1) \times (N_y + 1)$ mesh, we have

$$\alpha = \frac{1}{h^2} (2 - 2\cos(\pi h)) \approx \pi^2 \quad \text{and} \quad \beta = \frac{1}{h^2} (2 + 2\cos(\pi h)) \approx \frac{4}{h^2},$$

which, when substituted into (6.5), gives us

$$\rho_j \approx \frac{4}{h^2} \delta^{j-1}, \qquad j = 1, \ldots, n.$$

Parallel algorithm

We illustrate the parallelization of the ADI method on the strip-wise decomposition in Fig. 6.8. In this special case, we have only interior nodes and edge nodes, but no cross

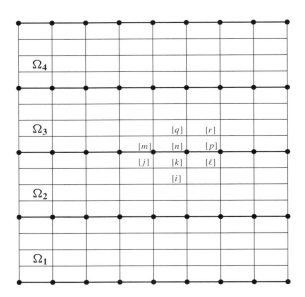

Figure 6.8. *Decomposition in four strips; • denotes an edge node.*

points. Additionally, there are no matrix connections between nodes from different edges. This implies the following block structure of the stiffness matrix:

$$K = \begin{pmatrix} K_E & K_{EI} \\ K_{IE} & K_I \end{pmatrix}$$

$$= \left(\begin{array}{ccccc|cccc} K_{E_{01}} & 0 & 0 & 0 & 0 & K_{E_{01}I_1} & 0 & 0 & 0 \\ 0 & K_{E_{12}} & 0 & 0 & 0 & K_{E_{12}I_1} & K_{E_{12}I_2} & 0 & 0 \\ 0 & 0 & K_{E_{23}} & 0 & 0 & 0 & K_{E_{23}I_2} & K_{E_{23}I_3} & 0 \\ 0 & 0 & 0 & K_{E_{34}} & 0 & 0 & 0 & K_{E_{34}I_3} & K_{E_{34}I_4} \\ 0 & 0 & 0 & 0 & K_{E_{40}} & 0 & 0 & 0 & K_{E_{40}I_4} \\ \hline K_{I_1E_{01}} & K_{I_1E_{12}} & 0 & 0 & 0 & K_{I_1} & 0 & 0 & 0 \\ 0 & K_{I_2E_{12}} & K_{I_2E_{23}} & 0 & 0 & 0 & K_{I_2} & 0 & 0 \\ 0 & 0 & K_{I_3E_{23}} & K_{I_3E_{34}} & 0 & 0 & 0 & K_{I_3} & 0 \\ 0 & 0 & 0 & K_{I_4E_{34}} & K_{I_4E_{40}} & 0 & 0 & 0 & K_{I_4} \end{array} \right).$$

with exactly that block structure for K_E that is required in point 3(a) in section 5.4. This means that formula (5.11a) can be applied with an accumulated matrix $\mathfrak{K}$ by setting

$$\mathfrak{M}_L := \begin{pmatrix} 0 & 0 \\ \mathfrak{K}_{IE} & 0 \end{pmatrix}, \qquad \mathfrak{M}_D := \begin{pmatrix} \mathfrak{K}_E & 0 \\ 0 & \mathfrak{K}_I \end{pmatrix}, \qquad \text{and } \mathfrak{M}_U := \begin{pmatrix} 0 & \mathfrak{K}_{EI} \\ 0 & 0 \end{pmatrix}.$$

Keeping in mind that $R_{I,i} = I_{I,i}$, we get

$$\underline{w} = \mathfrak{K} \cdot \underline{u} := \begin{pmatrix} \mathfrak{K}_E \cdot \underline{u}_E & + & \sum_{i=1}^{P} A_i^T \mathfrak{K}_{EI,i} \cdot \underline{u}_{I,i} \\ \mathfrak{K}_{IE} \cdot \underline{u}_E & + & \mathfrak{K}_I \cdot \underline{u}_I \end{pmatrix}. \tag{6.6}$$

Hence, we need only one communication step in the matrix-vector multiplication. The great advantage of using an accumulated matrix in ADI is that we only need one storage scheme for all the matrices in Algorithm 6.15. The specialties of

$$\mathfrak{V} = \begin{pmatrix} \mathfrak{K}_{E,y} & \mathfrak{K}_{IE} \\ \mathfrak{K}_{IE} & \mathfrak{K}_{I,y} \end{pmatrix} \quad \text{and} \quad \mathfrak{H} = \begin{pmatrix} \mathfrak{K}_{E,x} & 0 \\ 0 & \mathfrak{K}_{I,x} \end{pmatrix}$$

consist of a diagonal matrix $\mathfrak{K}_{E,y}$ and parallel invertible matrices $\mathfrak{K}_{E,x}$, $\mathfrak{K}_{I,x}$, $\mathfrak{K}_{I,y}$, $\mathfrak{T}_x :=$ $\mathfrak{H}+\rho_{k+1}\mathfrak{I}$. Therefore we can write the whole parallel ADI algorithm in terms of accumulated vectors and matrices.

Choose $\underline{u}^0$

$$\underline{\mathfrak{r}} := \begin{pmatrix} \underline{\mathfrak{f}}_E & - & \mathfrak{K}_E \cdot \underline{u}_E^0 & - & \sum_{i=1}^{P} A_i^T \mathfrak{K}_{EI,i} \cdot \underline{u}_{I,i}^0 \\ \underline{\mathfrak{f}}_I & - & \mathfrak{K}_{IE} \cdot \underline{u}_E^0 & - & \mathfrak{K}_I \cdot \underline{u}_I^0 \end{pmatrix}$$

$\sigma := \sigma_0 := (\underline{\mathfrak{r}}, R^{-1}\underline{\mathfrak{r}})$

$k := 0$

while $\quad \sigma > \text{tolerance}^2 \cdot \sigma_0 \quad$ do

$\qquad \mathfrak{T}_x := \mathfrak{H} + \rho_{k+1}\mathfrak{I}$

$$\underline{u}^{k*} = \begin{pmatrix} \mathfrak{T}_{E,x}^{-1}\left[\underline{\mathfrak{f}}_E & + & \rho_{k+1}\underline{u}_E^k & - & \mathfrak{K}_{E,y} \cdot \underline{u}_E^k & - & \sum_{i=1}^{P} A_i^T \mathfrak{K}_{EI,i} \cdot \underline{u}_{I,i}^k\right] \\ \mathfrak{T}_{I,x}^{-1}\left[\underline{\mathfrak{f}}_I & + & \rho_{k+1}\underline{u}_I^k & - & \mathfrak{K}_{IE} \cdot \underline{u}_E^k & - & \mathfrak{K}_{I,y} \cdot \underline{u}_I^k\right] \end{pmatrix}$$

$\qquad \mathfrak{T}_y := \mathfrak{V} + \rho_{k+1}\mathfrak{I}$

$$\underline{\mathfrak{g}} := \begin{pmatrix} \underline{\mathfrak{f}}_E & + & \rho_{k+1}\underline{u}_E^{k*} & - & \mathfrak{K}_{E,x} \cdot \underline{u}_E^{k*} \\ \underline{\mathfrak{f}}_I & + & \rho_{k+1}\underline{u}_I^{k*} & - & \mathfrak{K}_{I,x} \cdot \underline{u}_I^{k*} \end{pmatrix}$$

$\qquad$ Solve $\mathfrak{T}_y\underline{u}^{k+1} = \underline{\mathfrak{g}}$

$$\underline{\mathfrak{r}} := \begin{pmatrix} \underline{\mathfrak{f}}_E & - & \mathfrak{K}_E \cdot \underline{u}_E^{k+1} & - & \sum_{i=1}^{P} A_i^T \mathfrak{K}_{EI,i} \cdot \underline{u}_{I,i}^{k+1} \\ \underline{\mathfrak{f}}_I & - & \mathfrak{K}_{IE} \cdot \underline{u}_E^{k+1} & - & \mathfrak{K}_I \cdot \underline{u}_I^{k+1} \end{pmatrix}$$

$\qquad \sigma := (\underline{\mathfrak{r}}, R^{-1}\underline{\mathfrak{r}})$

$\qquad k := k + 1$

end

Algorithm 6.16: Parallel ADI in two dimensions: first try.

Algorithm 6.16 has the disadvantage of two communication steps in matrix-vector multiplications per iteration. This can be reduced to one communication step because the involved subvectors and submatrices are identical. We introduce the auxiliary vector $\underline{v}$ for this purpose and rewrite the algorithm. Note that $\underline{v}$ contains parts of the residual $\underline{\mathfrak{r}}$ and that it is related to $\underline{u}^{k+1}$. Hence, $\underline{v}$ can be reused in the next iteration to reduce the computational cost. The inversions $\mathfrak{T}_{E,x}^{-1}$ and $\mathfrak{T}_{I,x}^{-1}$ in Algorithm 6.16 and the inversion $\mathfrak{T}_x^{-1}$ in Algorithm 6.17 only use a sequential solver for tridiagonal band matrices along the lines in the x direction [94]. Hence, there is no communication at all.

On the other hand, a parallel solver for the tridiagonal system $\mathfrak{T}_y\underline{u}^{k+1} = \underline{\mathfrak{g}}$ in the y direction is needed. For simplicity we use an adapted version of Algorithm 6.14 for this

Choose $\underline{u}^0$

$$\underline{v} \quad := \quad \begin{pmatrix} \underline{f}_E & - & \sum_{i=1}^{P} A_i^T \mathfrak{K}_{EI,i} \cdot \underline{u}_{I,i}^0 \\ \underline{f}_I & - & \mathfrak{K}_{IE} \cdot \underline{u}_E^0 \end{pmatrix}$$

$$\underline{r} \quad := \quad \underline{v} - \begin{pmatrix} \mathfrak{K}_E & 0 \\ 0 & \mathfrak{K}_I \end{pmatrix} \cdot \underline{u}^0$$

$$\sigma \quad := \quad \sigma_0 := (\underline{r}, R^{-1}\underline{r})$$

$$k \quad := \quad 0$$

while $\quad \sigma > \text{tolerance}^2 \cdot \sigma_0 \quad$ do

$$\mathfrak{T}_x \quad := \quad \mathfrak{H} + \rho_{k+1}\mathfrak{J}$$

$$\underline{u}^{k*} \quad := \quad \mathfrak{T}_x^{-1} \cdot \left[\underline{v} + \rho_{k+1}\underline{u}^k - \begin{pmatrix} \mathfrak{K}_{E,y} & 0 \\ 0 & \mathfrak{K}_{I,y} \end{pmatrix} \cdot \underline{u}^k \right]$$

$$\mathfrak{T}_y \quad := \quad \mathfrak{V} + \rho_{k+1}\mathfrak{J}$$

$$\underline{g} \quad := \quad \underline{f} + \rho_{k+1}\underline{u}^{k*} - \begin{pmatrix} \mathfrak{K}_{E,x} & 0 \\ 0 & \mathfrak{K}_{I,x} \end{pmatrix} \cdot \underline{u}^{k*}$$

Solve $\mathfrak{T}_y \underline{u}^{k+1} = \underline{g}$

$$\underline{v} \quad := \quad \begin{pmatrix} \underline{f}_E & - & \sum_{i=1}^{P} A_i^T \mathfrak{K}_{EI,i} \cdot \underline{u}_{I,i}^{k+1} \\ \underline{f}_I & - & \mathfrak{K}_{IE} \cdot \underline{u}_E^{k+1} \end{pmatrix}$$

$$\underline{r} \quad := \quad \underline{v} - \begin{pmatrix} \mathfrak{K}_E & 0 \\ 0 & \mathfrak{K}_I \end{pmatrix} \cdot \underline{u}^{k+1}$$

$$\sigma \quad := \quad (\underline{r}, R^{-1}\underline{r})$$

$$k \quad := \quad k+1$$

end

Algorithm 6.17: Parallel ADI in two dimensions: final version.

purpose. It is obvious that $\mathfrak{T}_{y,E}$ is a diagonal matrix and $\mathfrak{T}_{y,I}$ is equivalent to a block diagonal matrix with tridiagonal blocks. In this case, we can solve systems with these matrices by means of a direct solver at low cost. This changes

$$\widehat{\underline{u}}_E^{\ell+1} := \widehat{\underline{u}}_E^\ell + \mathfrak{T}_{y,E}^{-1} \cdot \left(\underline{g}_E - \mathfrak{T}_{y,E} \cdot \widehat{\underline{u}}_E^\ell - \sum_{s=1}^{P} A_s^T \mathfrak{T}_{y,EI,s} \cdot \widehat{\underline{u}}_{I,s}^\ell \right)$$

from the straightforward Gauss–Seidel implementation into

$$\widehat{\underline{u}}_E^{\ell+1} := \mathfrak{T}_{y,E}^{-1} \cdot \left(\underline{g}_E - \sum_{s=1}^{P} A_s^T \mathfrak{T}_{y,EI,s} \cdot \widehat{\underline{u}}_{I,s}^\ell \right)$$

and is applied in a block iteration for solving the system. Again we use an auxiliary variable $\widehat{\underline{v}}$ to save communication steps (see Algorithm 6.18).

$$\widehat{\underline{u}}^0 \quad := \underline{u}^{k*}, \qquad \ell := 0$$

$$\widehat{\underline{v}}_E \quad := \underline{g}_E - \sum_{s=1}^{P} A_s^T \mathfrak{T}_{y,EI,s} \cdot \widehat{\underline{u}}_{I,s}^0$$

$$\text{while} \quad \sigma > \text{tolerance}^2 \cdot \sigma_0 \quad \text{do}$$

$$\widehat{\underline{u}}_E^{\ell+1} \quad := \mathfrak{T}_{y,E}^{-1} \cdot \widehat{\underline{v}}_E$$

$$\widehat{\underline{v}}_I \quad := \underline{g}_I - \mathfrak{T}_{y,IE} \cdot \widehat{\underline{u}}_E^{\ell+1}$$

$$\widehat{\underline{u}}_I^{\ell+1} \quad := \mathfrak{T}_{y,I}^{-1} \cdot \widehat{\underline{v}}_I$$

$$\widehat{\underline{v}}_E \quad := \underline{g}_E - \sum_{s=1}^{P} A_s^T \mathfrak{T}_{y,EI,s} \cdot \widehat{\underline{u}}_{I,s}^{\ell+1}$$

$$\widehat{\underline{r}} \quad := \widehat{\underline{v}} - \begin{pmatrix} \mathfrak{T}_{y,E} & 0 \\ 0 & \mathfrak{T}_{y,I} \end{pmatrix} \cdot \widehat{\underline{u}}^{\ell+1}$$

$$\sigma \quad := (\widehat{\underline{r}}, R^{-1}\widehat{\underline{r}})$$

$$\ell \quad := \ell + 1$$

$$\text{end}$$

$$\underline{u}^{k+1} \quad := \widehat{\underline{u}}^{\ell}$$

Algorithm 6.18: Gauss–Seidel iteration for solving $\mathfrak{T}_y \underline{u}^{k+1} = \underline{g}$.

6.3 Roughers

As noted in the introduction to section 6.2, not all iterative methods are smoothers. Some are roughers.

A rougher differs from a smoother in how individual error components are treated. While the error norm of a rougher should decrease each iteration, some of the individual components may increase in magnitude. Roughers differ from proper smoothers (where no error component increases in magnitude each iteration) in how the error components behave during the iterative procedure.

6.3.1 CG method

Sequential algorithm

If matrix K in (6.1) is SPD (($K\underline{v}, \underline{v}) > 0 \ \forall \ v \neq 0$), then we may use the conjugate gradient (CG) method for solving (6.1). Algorithm 6.19 presents this method in combination with an SPD preconditioner C.

Parallelized CG

Investigating the operations in the classical CG (Algorithm 6.19 with $C = I$) with respect to the results of sections 5.3–5.4, we use the following *strategy for parallelization*. One objective therein is the minimization of communication steps.

First, we store matrix K as distributed K (5.1) because otherwise we have to assume restrictions on the matrix pattern. Second, vectors $\underline{s}, \underline{u} \to \underline{s}, \underline{u}$ are accumulated (Type I) stored and so the appropriate matrix-vector product results in a Type II vector without any

$$
\begin{aligned}
&\text{Choose } \underline{u}^0 \\
&\underline{r} \;:=\; \underline{f} - K \cdot \underline{u}^0 \\
&\underline{w} \;:=\; C^{-1} \cdot \underline{r} \\
&\underline{s} \;:=\; \underline{w} \\
&\sigma \;:=\; \sigma_{\text{old}} \;:=\; \sigma_0 \;:=\; (\underline{w}, \underline{r}) \\
&\text{repeat} \\
&\qquad \underline{v} \quad:=\; K \cdot \underline{s} \\
&\qquad \alpha \quad:=\; \sigma/(\underline{s}, \underline{v}) \\
&\qquad \underline{u} \quad:=\; \underline{u} + \alpha \cdot \underline{s} \\
&\qquad \underline{r} \quad:=\; \underline{r} - \alpha \cdot \underline{v} \\
&\qquad \underline{w} \quad:=\; C^{-1} \cdot \underline{r} \\
&\qquad \sigma \quad:=\; (\underline{w}, \underline{r}) \\
&\qquad \beta \quad:=\; \sigma/\sigma_{\text{old}} \\
&\qquad \sigma_{\text{old}} \;:=\; \sigma \\
&\qquad \underline{s} \quad:=\; \underline{w} + \beta \cdot \underline{s} \\
&\text{until} \quad \sqrt{\sigma/\sigma_0} \;<\; \text{tolerance}
\end{aligned}
$$

Algorithm 6.19: Sequential CG with preconditioning.

$$
\begin{aligned}
&\text{Choose } \underline{\mathfrak{u}}^0 \\
&\underline{\mathfrak{r}} \;:=\; \underline{\mathfrak{f}} - \mathsf{K} \cdot \underline{\mathfrak{u}}^0 \\
&\underline{\mathfrak{w}} \;:=\; \sum_{j=1}^{P} A_j^T \underline{\mathfrak{r}}_j \\
&\underline{\mathfrak{s}} \;:=\; \underline{\mathfrak{w}} \\
&\sigma \;:=\; \sigma_{\text{old}} \;:=\; \sigma_0 \;:=\; (\underline{\mathfrak{w}}, \underline{\mathfrak{r}}) \\
&\text{repeat} \\
&\qquad \underline{\mathfrak{v}} \quad:=\; \mathsf{K} \cdot \underline{\mathfrak{s}} \\
&\qquad \alpha \quad:=\; \sigma/(\underline{\mathfrak{s}}, \underline{\mathfrak{v}}) \\
&\qquad \underline{\mathfrak{u}} \quad:=\; \underline{\mathfrak{u}} + \alpha \cdot \underline{\mathfrak{s}} \\
&\qquad \underline{\mathfrak{r}} \quad:=\; \underline{\mathfrak{r}} - \alpha \cdot \underline{\mathfrak{v}} \\
&\qquad \underline{\mathfrak{w}} \quad:=\; \sum_{j=1}^{P} A_j^T \underline{\mathfrak{r}}_j \\
&\qquad \sigma \quad:=\; (\underline{\mathfrak{w}}, \underline{\mathfrak{r}}) \\
&\qquad \beta \quad:=\; \sigma/\sigma_{\text{old}} \\
&\qquad \sigma_{\text{old}} \;:=\; \sigma \\
&\qquad \underline{\mathfrak{s}} \quad:=\; \underline{\mathfrak{w}} + \beta \cdot \underline{\mathfrak{s}} \\
&\text{until} \quad \sqrt{\sigma/\sigma_0} \;<\; \text{tolerance}
\end{aligned}
$$

Algorithm 6.20: Parallelized CG.

communication (5.8). This leads to the obvious conclusion that vectors $\underline{v}, \underline{r}, \underline{f} \rightarrow \underline{\mathfrak{v}}, \underline{\mathfrak{r}}, \underline{\mathfrak{f}}$ should be stored distributed (Type II) and are never to be accumulated during the iteration. Additionally, if $\underline{w} \rightarrow \underline{\mathfrak{w}}$ is an accumulated vector, then all DAXPY operations require no communication. Due to the different types of vectors used in the inner products, these operations require only a very small amount of communication (5.7).

Note that choosing $C = I$ in Algorithm 6.19 results in the setting of $\underline{w} := \underline{r}$, which includes one change of the vector type via accumulation (5.5). Here we need communication between all the processes sharing the proper data. The resulting parallel CG in Algorithm 6.20 requires two ALL_REDUCE operations with one real number and one vector accumulation.

6.3.2 GMRES solver

Sequential algorithm

To solve a system of linear equations with a nonsymmetric matrix K ($K \neq K^T$), we can use the general conjugate residual method [93]. If K is additionally not positive definite, then a variety of methods such as generalized minimum residual (GMRES), quasi-minimal residual (QMR), Bi-CG, and biconjugate gradient stabilized (BICSTAB), can be used instead.

$$
\begin{aligned}
&\text{Choose } \underline{u}^0 \\
&\underline{r} \quad := \underline{f} - K \cdot \underline{u}^0 \\
&\underline{w}^1 := C^{-1} \cdot \underline{r} \\
&z_1 \quad := (\underline{w}^1, \underline{r}) \\
&k \quad := 0 \\
&\text{repeat} \quad k := k+1 \\
&\qquad\qquad \underline{r} \qquad := K \cdot \underline{w}^k \\
&\qquad\qquad \text{for } i := 1 \text{ to } k \text{ do} \\
&\qquad\qquad\qquad\qquad h_{i,k} := (\underline{w}^i, \underline{r}) \\
&\qquad\qquad\qquad\qquad \underline{r} \quad := \underline{r} - h_{i,k} \cdot \underline{w}^i \\
&\qquad\qquad \text{end} \\
&\qquad\qquad \underline{w}^{k+1} := C^{-1} \cdot \underline{r} \\
&\qquad\qquad h_{k+1,k} := (\underline{w}^{k+1}, \underline{r}) \\
&\qquad\qquad \underline{w}^{k+1} := \underline{w}^{k+1}/h_{k+1,k} \\
&\qquad\qquad \text{for } i := 1 \text{ to } k-1 \text{ do} \begin{pmatrix} h_{i,k} \\ h_{i+1,k} \end{pmatrix} := \begin{pmatrix} c_{i+1} & s_{i+1} \\ s_{i+1} & -c_{i+1} \end{pmatrix} \cdot \begin{pmatrix} h_{i,k} \\ h_{i+1,k} \end{pmatrix} \\
&\qquad\qquad \alpha \qquad := \sqrt{h_{k,k}^2 + h_{k+1,k}^2} \\
&\qquad\qquad s_{k+1} := h_{k+1,k}/\alpha \; ; \; c_{k+1} := h_{k,k}/\alpha \; ; \; h_{k,k} := \alpha \\
&\qquad\qquad z_{k+1} := s_{k+1}z_k \; ; \; z_k := c_{k+1}z_k \\
&\text{until} \quad |z_{k+1}/z_1| < \text{tolerance} \\
&z_k \quad := z_k/h_{k,k} \\
&\text{for } i := k-1 \text{ down to 1 do} \quad z_i := \left(z_i - \sum_{j=i+1}^{k} h_{i,j}z_j \right)/h_{i,i} \\
&\underline{u}^k := \underline{u}^0 + \sum_{i=1}^{k} z_i \cdot \underline{w}^j
\end{aligned}
$$

Algorithm 6.21: Sequential GMRES with preconditioning.

Let us consider the GMRES method in Algorithm 6.21. The number of stored vectors $\underline{w}^k$ and entries in matrix $H = \{h_{i,j}\}_{i,j=\overline{1,k}}$ increases with the number of iterations k and may require excessive amounts of memory. To solve this problem, the REPEAT loop is stopped after m iterations and restarted with $\underline{u}^m$ as the initial guess. This shortening of the cycle leads to unstable behavior of the method. For preconditioning techniques, see section 6.2.2.

Parallel GMRES

We have to distinguish in Algorithm 6.21 not only between functionals f, $\underline{r}$ and state variables $\underline{u}$, $\underline{w}^k$, but also between vectors $\{z_i\}_{i=\overline{1,k+1}}$, $\{s_i\}_{i=\overline{1,k+1}}$, $\{c_i\}_{i=\overline{1,k+1}}$ and matrix $\{h_{i,j}\}_{i,j=\overline{1,k}}$. None of these fits into that scheme. Therefore, we handle them simply as scalar values in the parallelization.

$$
\begin{aligned}
&\text{Choose } \underline{u}^0 \\
&\underline{r} \quad := \underline{f} - \mathrm{K} \cdot \underline{u}^0 \\
&\underline{w}^1 := \sum_{s=1}^{P} A_s^T \underline{r}_s \\
&z_1 \quad := (\underline{w}^1, \underline{r}) \\
&k \quad := 0 \\
&\text{repeat} \quad k := k + 1 \\
&\qquad \underline{r} \qquad := \mathrm{K} \cdot \underline{w}^k \\
&\qquad \text{for } i := 1 \text{ to } k \text{ do} \\
&\qquad\qquad\qquad h_{i,k} := (\underline{w}^i, \underline{r}) \\
&\qquad\qquad\qquad \underline{r} \quad := \underline{r} - h_{i,k} \cdot R^{-1} \cdot \underline{w}^i \\
&\qquad \text{end} \\
&\qquad \underline{w}^{k+1} := \sum_{s=1}^{P} A_s^T \underline{r}_s \\
&\qquad h_{k+1,k} := (\underline{w}^{k+1}, \underline{r}) \\
&\qquad \underline{w}^{k+1} := \underline{w}^{k+1} / h_{k+1,k} \\
&\qquad \text{for } i := 1 \text{ to } k-1 \text{ do} \quad \begin{pmatrix} h_{i,k} \\ h_{i+1,k} \end{pmatrix} := \begin{pmatrix} c_{i+1} & s_{i+1} \\ s_{i+1} & -c_{i+1} \end{pmatrix} \cdot \begin{pmatrix} h_{i,k} \\ h_{i+1,k} \end{pmatrix} \\
&\qquad \alpha \qquad := \sqrt{h_{k,k}^2 + h_{k+1,k}^2} \\
&\qquad s_{k+1} := h_{k+1,k}/\alpha \; ; \; c_{k+1} := h_{k,k}/\alpha \; ; \; h_{k,k} := \alpha \\
&\qquad z_{k+1} := s_{k+1} z_k \; ; \; z_k := c_{k+1} z_k \\
&\text{until} \quad |z_{k+1}/z_1| < \text{tolerance} \\
&z_k := z_k / h_{k,k} \\
&\text{for } i := k-1 \text{ down to } 1 \text{ do} \quad z_i := \left(z_i - \sum_{j=i+1}^{k} h_{i,j} z_j \right) / h_{i,i} \\
&\underline{u}^k := \underline{u}^0 + \sum_{i=1}^{k} z_i \cdot \underline{w}^j
\end{aligned}
$$

Algorithm 6.22: Parallel GMRES—no restart.

Our *strategy for the parallelization* of GMRES without preconditioning ($C = I$) is the following: vectors f, $\underline{r} \to f$, $\underline{r}$ are stored distributed and vectors $\underline{u}$, $\underline{w}^k \to \underline{u}$, $\underline{w}^k$ are stored accumulated. We store the scalar values $\{z_i\}_{i=\overline{1,k+1}}$, $\{s_i\}_{i=\overline{1,k+1}}$, $\{c_i\}_{i=\overline{1,k+1}}$, and

$\{h_{i,j}\}_{i,j=\overline{1,k}}$ redundantly on each processor. Then all inner products require only a minimum of communication (5.7) and the matrix-vector operation requires no communication at all (5.8). Setting $\underline{\mathfrak{w}} := \underline{r}$ requires one change of the vector type via accumulation (5.5). Here we need communication between all processes sharing the proper data. All operations on scalar values c_i, s_i, z_i, and $h_{i,j}$ are performed locally on each processor. These operations are redundant. *Nearly all* DAXPY operations use the same vector types (and scalar values) *excluding* the operation

$$\underline{r} := \underline{r} - h_{i,k} \cdot \underline{\mathfrak{w}}^i,$$

which combines different vector types. But the necessary change of an accumulated vector $\underline{\mathfrak{w}}^i$ into a distributed one can be done without any communication (5.6). Therefore, no DAXPY operation requires communication.

The *k*th iteration of Algorithm 6.22 includes $(k + 1)$ ALL_REDUCE operations with one real number and one vector accumulation. Due to changing the type of vector $\underline{\mathfrak{w}}^i$ we have to perform $k \cdot n$ additional multiplications (n is the length of vector $\underline{\mathfrak{w}}$).

A parallelization of GMRES(m) with a restart after m iterations can be done in the same way as above.

6.3.3   BICGSTAB solver

Sequential algorithm

One method for solving a system of linear equations with a nonsymmetric and not positive definite matrix K is BICGSTAB in Algorithm 6.23.

Choose initial guess $\underline{u}$
$\underline{r} := \underline{f} - K \cdot \underline{u}$
Choose $\underline{q}$ such that $(\underline{q}, \underline{r}) \neq 0$
Set $\varrho := 1$, $\alpha := 1$, $\omega := 1$, $\underline{v} := 0$, $\underline{p} := 0$
while $(\underline{r}, \underline{r}) < \varepsilon^2$ do
$\quad \varrho_{old} := \varrho$
$\quad \varrho \quad := (\underline{q}, \underline{r})$
$\quad \beta \quad := \varrho/\varrho_{old} \cdot \alpha/\omega$
$\quad \underline{p} \quad := \underline{r} + \beta \cdot (\underline{p} - \omega \cdot \underline{v})$
$\quad \underline{v} \quad := K \cdot \underline{p}$
$\quad \alpha \quad := \varrho/(\underline{q}, \underline{v})$
$\quad \underline{s} \quad := \underline{r} - \alpha \cdot \underline{v}$
$\quad \underline{t} \quad := K \cdot \underline{s}$
$\quad \omega \quad := (\underline{t}, \underline{s})/(\underline{t}, \underline{t})$
$\quad \underline{u} \quad := \underline{u} + \alpha \cdot \underline{p} + \omega \cdot \underline{s}$
$\quad \underline{r} \quad := \underline{s} - \omega \cdot \underline{t}$

Algorithm 6.23: Sequential BICGSTAB.

Parallel BICGSTAB

If we choose the same parallelization strategy as in section 6.3.2, then the parallel BICGSTAB in Algorithm 6.24 requires three type conversions with communication per iteration and two additional vectors. This can be reduced by using mostly accumulated vectors in the parallel code.

$$
\begin{aligned}
&\text{Choose initial guess } \underline{u} \\
&\underline{r} := \sum_{s=1}^{P} A_s^T \left(\underline{f}_s - K_s \cdot \underline{u}_s \right) \\
&\text{Choose } \underline{q} \text{ such that } (\underline{q}, \underline{r}) \neq 0 \\
&\text{Set } \varrho := 1,\ \alpha := 1,\ \omega := 1,\ \underline{v} := 0,\ \underline{p} := 0 \\
&\text{while } (\underline{r}, R^{-1}\underline{r}) < \varepsilon^2 \quad \text{do} \\
&\qquad \varrho_{old} := \varrho \\
&\qquad \varrho \quad := (\underline{q}, \underline{r}) \\
&\qquad \beta \quad := \varrho / \varrho_{old} \cdot \alpha / \omega \\
&\qquad \underline{p} \quad := \underline{r} + \beta \cdot (\underline{p} - \omega \cdot \underline{v}) \\
&\qquad \underline{v} \quad := \sum_{s=1}^{P} A_s^T K_s \cdot \underline{p}_s \\
&\qquad \alpha \quad := \varrho / (\underline{q}, \underline{v}) \\
&\qquad \underline{s} \quad := \underline{r} - \alpha \cdot \underline{v} \\
&\qquad \underline{t} \quad := K \cdot \underline{s} \\
&\qquad \widehat{\underline{t}} \quad := \sum_{s=1}^{P} A_s^T \underline{t}_s \\
&\qquad \omega \quad := (\underline{t}, \underline{s}) / (\underline{t}, \widehat{\underline{t}}) \\
&\qquad \underline{u} \quad := \underline{u} + \alpha \cdot \underline{p} + \omega \cdot \underline{s} \\
&\qquad \underline{r} \quad := \underline{s} - \omega \cdot \widehat{\underline{t}}
\end{aligned}
$$

Algorithm 6.24: Parallel BICGSTAB.

6.4   Preconditioners

The parallel versions of classical iterative schemes listed in sections 6.1.2 to 6.2.3 can be used as preconditioners in sections 6.3.1 and 6.3.2. Additionally, the symmetric multigrid iteration in Chapter 7 also can be used as a preconditioner [68, 67].

Exercises

We want to solve the Laplace problem using the weak formulation with homogeneous Dirichlet boundary conditions (BCs) on the discretization given in Fig. 5.4.

E6.1. You will find a sequential version of the Jacobi solver in the directory *Solution/jacseqc* with the following functions in addition to the functions in the Exercises for Chapter 5:

GetMatrix(nx, ny, xl, xr, yb, yt, sk, id, ik, f)

in which matrix K and right-hand side f are calculated using the function **FunctF(x, y)** for describing $\mathbf{f}(x)$. Note that only coordinates and element connectivities are related to the rectangular domain. All other parts in this routine are written for general 2D domains.

The Dirichlet BCs are set in **SetU(nx, ny, u)**. Alternatively, we could use **FunctU(x,y)**. These BCs are applied in

ApplyDirichletBC(nx, ny, neigh, u, sk, id, ik, f)

via a penalty method.

The solver itself is implemented in

JacobiSolve(nx, ny, sk, id, ik, f, u)

and uses

GetDiag(nx, ny, sk, id, d)

to get the diagonal from matrix K. Matrix-vector multiplication is implemented in

CrsMult(iza, ize, w, u, id, ik, sk, alfa) .

A vector $\underline{u}$ can be saved in a file named *name* by calling

SaveVector(name, u, nx, ny, xl, xr, yb, yt, ierr)

such that `gnuplot name` can be used to visualize that vector.

Implement a parallel ω-Jacobi solver based on the sequential code.

E6.2. Implement a parallel CG without preconditioning.

E6.3. Implement a parallel CG with diagonal preconditioning.

Chapter 7

Multigrid Methods

Theory attracts practice as the magnet attracts iron.
—Johann C. F. Gauss (1777–1855)

7.1 Multigrid methods

We assume that there exists a series of regular (finite element) meshes/grids $\{\mathfrak{T}_q\}_{q=1}^{\ell}$, where the *finer grid* $\mathfrak{T}_{q+1}$ was derived from the *coarser grid* $\mathfrak{T}_q$. The simplest case in two dimensions is to subdivide all triangles into four congruent ones, i.e., by connecting the bisection points of the three edges.

We assume that the grids are nested: $\mathfrak{T}_1 \subset \mathfrak{T}_2 \subset \cdots \subset \mathfrak{T}_\ell$. There are convergence results for nonnested triangulations (and more general griddings and discretizations) [27, 28], but we do not consider them. This simplifies the theory considerably [5, 58, 114].

Discretizing the given differential equation on each grid $\mathfrak{T}_q$ results in a series of systems of equations

$$K_q \underline{u}_q = \underline{f}_q, \tag{7.1}$$

with the symmetric positive definite (SPD) sparse stiffness matrices K_q (see Chapter 4 for finite element meshes).

When analyzing the nonoptimal convergence behavior of iterative methods presented in sections 6.2.1 and 6.2.2 by means of representing the error $\underline{e} := \underline{u}_q^* - \underline{u}_q^k$ as the sum of the eigenfrequencies of the matrix K, we achieve the following results (see also the nicely written multigrid introduction [16]). First, those parts of the error belonging to high eigenfrequencies are reduced very quickly. Second, low-frequency parts of the error yield to the slow convergence behavior of classical iterative solvers.

We need an inexpensive method to reduce the low-frequency errors e_{low} without losing the good convergence behavior for the high-frequency errors e_{high}. This leads directly to the *two-grid idea:*

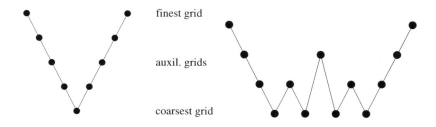

finest grid

auxil. grids

coarsest grid

Figure 7.1. *V- and W-cycles.*

Reduce e_{high} on grid $\mathfrak{T}_\ell$ (smoothing), project the remaining error onto coarse grid $\mathfrak{T}_{\ell-1}$, and solve there exactly. Interpolate the coarse grid solution (called the defect correction) back to the fine grid $\mathfrak{T}_\ell$ and add it to the approximation obtained after smoothing.

Usually, additional smoothing reduces the high-frequency errors once more on $\mathfrak{T}_\ell$. The two-grid method is repeated until convergence is reached. We defer until **Coarse grid solvers** in section 7.2.2 the proper definition of *solve exactly* but, until then, assume a direct solver is adequate.

The *multigrid idea* substitutes the exact solver in the defect correction on the coarse grid $\mathfrak{T}_{\ell-1}$ recursively with another two-grid method until the coarsest grid $\mathfrak{T}_1$ is reached. There we solve the remaining small system of equations exactly by a direct method.

7.2   The multigrid algorithm

7.2.1   Sequential algorithm

For notational purposes we have to introduce the following vectors and matrices:

I_q^{q-1} restriction/projection operator for mapping fine grid data onto the coarse grid;

I_{q-1}^q prolongation/interpolation operator for mapping coarse grid data onto the fine grid;

$S_{pre}^{\nu_{pre}}$ presmoothing operator reducing e_{high}, applied ν_{pre} times;

$S_{post}^{\nu_{post}}$ postsmoothing operator reducing e_{high}, applied ν_{post} times;

$\underline{d}_q$ defect on qth grid;

$\underline{w}_q$ correction on qth grid;

MGM$^\gamma$ recursive multigrid procedure, applied γ times ($\gamma = 1$: V-cycle, $\gamma = 2$: W-cycle) (Fig. 7.1).

We choose one or more iterative methods from sections 6.2.1 and 6.2.2 as the presmoother or postsmoother in Algorithm 7.1. Note that they can be different. If we use multigrid as preconditioner in a conjugate gradient (CG) algorithm, then the proper multigrid operator

if $(q == 1)$ then
  Solve $K_1 \underline{u}_1 = \underline{f}_1$ exactly   Coarse grid solver
else

$\widehat{\underline{u}}_q$	$:= S^{\nu_{pre}}_{pre}(K_q, \underline{u}_q, \underline{f}_q)$	Presmoothing
$\underline{d}_q$	$:= \underline{f}_q - K_q \cdot \widehat{\underline{u}}_q$	Defect calculation
$\underline{d}_{q-1}$	$:= I^{q-1}_q \cdot \underline{d}_q$	Restriction of defect (new r.h.s.)
$\underline{w}^0_{q-1}$	$:= 0$	Initial guess
$\underline{w}_{q-1}$	$:= \text{MGM}^\gamma(K_{q-1}, \underline{w}^0_{q-1}, \underline{d}_{q-1}, q-1)$	Defect system
$\underline{w}_q$	$:= I^q_{q-1} \cdot \underline{w}_{q-1}$	Interpolation of correction
$\widetilde{\underline{u}}_q$	$:= \widehat{\underline{u}}_q + \underline{w}_q$	Add correction
$\underline{u}_q$	$:= S^{\nu_{post}}_{post}(K_q, \widetilde{\underline{u}}_q, \underline{f}_q)$	Postsmoothing

endif

Algorithm 7.1: Sequential multigrid: $\text{MGM}(K_q, \underline{u}_q, \underline{f}_q, q)$.

must be SPD. In this case we must choose the grid transfer operators, i.e., interpolation and restriction, as $I^{q-1}_q = (I^q_{q-1})^T$ and the iteration operator of the postsmoother has to be adjoint to the iteration operator of the presmoother with respect to the energy product $(\cdot, \cdot)_{K_q}$. Additionally, the number of smoothing steps has to be the same, i.e., $\nu_{pre} = \nu_{post}$, and the postsmoother as well as the presmoother has to start with an initial guess equal to zero [68, 67]. The defect system on the coarsest grid is solved exactly by a direct method or it can be solved by an iterative method with a symmetric iteration operator, e.g., ω-Jacobi.

7.2.2 Parallel components of multigrid

It is obvious that we can parallelize the whole multigrid algorithm if we can parallelize each single component of it, i.e., interpolation, restriction, smoothing, and the coarse grid solver. If we use the nonoverlapping element distribution from section 5.2 on the coarsest grid $\mathfrak{T}_1$, then this distribution is kept on all finer grids.

  Once again, we store the matrices K_q as distributed ones. From the experiences of previous sections (especially section 6.3.1), the following classification of vectors from Algorithm 7.1 into the two parallel vector types seems to be advantageous:

- accumulated vectors: $\widehat{\underline{u}}_q, \widetilde{\underline{u}}_q, \widehat{\underline{w}}_q$;

- distributed vectors: $\underline{f}_q, \underline{d}_q$;

- not yet assigned to a vector type: $\underline{u}_q, \underline{w}_{q-1}, \underline{w}^0_{q-1}, \underline{d}_{q-1}$.

Interpolation

Investigating Fig. 7.2 and the proper interpolation resulting from the mesh refinement we can derive for the three sets of nodes V, E, I the following statements:

1. The interpolation $I^q_{V,q-1}$ within the set of cross points is the (potentially scaled) identity matrix I.

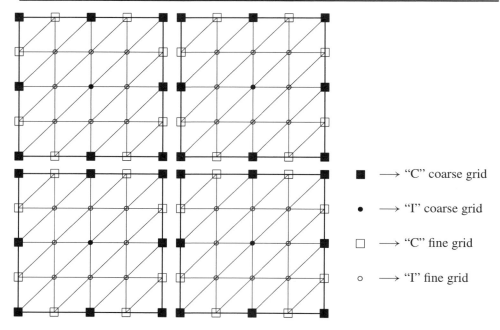

Figure 7.2. *Nonoverlapping element distribution on two nested grids.*

2. The complete set of crosspoints belongs to the coarse grid and therefore they are never interpolated by nodes belonging to E or I; i.e., $I_{VE,q-1}^q = I_{VI,q-1}^q = 0$.

3. The interpolation matrix on the edges breaks up into blocks and so we get
$I_{E,q-1}^q = \text{blockdiag}\{I_{E_j,q-1}^q\}_{i=\overline{1,\text{NumEdges}}}$.

4. Nodes on the edges are never interpolated by inner nodes; i.e., $I_{EI,q-1}^q = 0$.

5. Obviously, the interpolation matrix of the inner nodes (I) is also block diagonal; i.e.,
$I_{I,q-1}^q = \text{blockdiag}\{I_{I_s,q-1}^q\}_{s=\overline{1,P}}$.

As a consequence, the interpolation matrix I_{q-1}^q possesses a block structure in which the submatrices on the upper right are equal to 0. The structure of the matrix allows the application of (5.9) and we conclude that

- the interpolation matrix has to be accumulated—$\mathfrak{I}_{q-1}^q$—and

- the vector $\underline{\mathfrak{w}}_{q-1}$ is accumulated.

In principle it is also possible to define the interpolation matrix as distributed. But in that case an additional type conversion with communication has to be performed after the interpolation or in the following addition of the correction:

No communication in interpolation! :-)

Restriction

By choosing the restriction as transposed interpolation, i.e., $I_q^{q-1} = (I_{q-1}^q)^T$, we achieve a block triangular structure of the restriction matrix, as in (5.10), leading to

- an accumulated restriction matrix being accumulated—$\mathfrak{J}_q^{q-1}$—and

- distributed stored vectors $\underline{d}_{q-1}$ and $\underline{d}_q$.

Again, it is also possible to define the restriction matrix as distributed. In this case an additional type conversion with communication has to be performed on $\underline{d}_q$ before the restriction. Using injection as restriction does not change any of the previous statements:

No communication in restriction! :-)

Remark 7.1. Choosing $I_q^{q-1} = (I_{q-1}^q)^T$ has an odd side effect for the finite difference method (FDM) (recall the FDM from Chapter 2). The resulting K_{q-1} resembles a finite element method (FEM) matrix. Instead of maintaining a five-point stencil (see (2.4)), a nine-point stencil is created. A much more complicated system is needed for FDM to maintain the five-point stencil on all grids (see [27] and [28] for unifying approaches).

Smoothing

The smoothing operators S_{pre} and S_{post} may require extra input vectors. Iterative methods like ω-Jacobi, Gauss–Seidel, and ILU (sections 6.2.1 and 6.2.2) require as input an accumulated initial solution $\underline{u}^0$, a distributed stored right-hand side $\underline{f}$, and a distributed matrix K. The iterated solution $\underline{u}$ is accumulated in any case. Each iteration of a smoother requires at least one accumulation. The appropriate storage classes will be

- vector $\underline{u}_q$ is accumulated, $\underline{\widehat{u}}_q$ and $\underline{\widetilde{u}}_q$ are accumulated in any case;

- vector $\underline{w}_q^0$ is also accumulated.

Communication in smoothing. :-(

Coarse grid solver

The main problem in parallelizing a multigrid or multilevel algorithm consists of solving the coarse grid system exactly:

$$\sum_{s=1}^P A_{s,1}^T K_{s,1} A_{s,1} \underline{u}_1 = \sum_{s=1}^P A_{s,1}^T \underline{f}_{s,1}. \tag{7.2}$$

The following ideas are common:

1. *Direct solver—sequential*:

   The coarsest grid is so small that one processor $\mathbb{P}_0$ can store it additionally.
   - Accumulate K_1 (REDUCE).
   - Accumulate $\underline{f}_1$ (REDUCE).

- Processor $\mathbb{P}_0$ solves $\mathfrak{K}_1 \underline{u}_1 = \underline{f}_1$.
- Distribute $\underline{u}_1$ to all other processors (SCATTER).

This algorithm is purely sequential!

2. *Direct solver—parallel*:

   We store parts of the accumulated matrix on all processors and solve the coarse grid system by some direct parallel solver. Therein, load imbalances can appear.

3. *Iterative solver—parallel*:

   Here, no accumulation of K_1 and $\underline{f}_1$ is necessary, but in each iteration at least one accumulation must be performed (see sections 6.3.1 and 6.3.2).

   Caution: If CG-like methods are used as coarse grid solvers, then a symmetric multi-grid operator cannot be achieved. To preserve the symmetry one should combine, e.g., a Gauss–Seidel forward iteration with the appropriate backward iteration (symmetric successive overrelaxation (SSOR)) or one should use algebraic multigrid (AMG) [91, 16, 56] as a coarse grid solver.

The usually poor ratio between communication and arithmetic behavior of parallel machines normally results in poor parallel efficiency for the coarse grid solver. This can be observed when using a W-cycle on 32 or more processors, since the W-cycle works on the coarser grids more often than a V-cycle does.

One idea to overcome this bottleneck consists of the distribution of the coarse (or the coarser) grid on fewer than P processors. The disadvantage of this approach is the appearance of communication between certain grids ℓ_0 and $\ell_0 + 1$ in the intergrid transfer:

Communication in coarse grid solver. :-(

7.2.3   Parallel algorithm

---

if $(q == 1)$ then

    Solve $\displaystyle\sum_{s=1}^{P} A_{s,1}^T K_{s,1} A_{s,1}\, \underline{u}_1 \;=\; \sum_{s=1}^{P} A_{s,1}^T \underline{f}_{s,1}$ Coarse grid solver

else

$\begin{aligned}
\widehat{\underline{u}}_q &:= S_{\text{pre}}^{\nu_{\text{pre}}}(K_q, \underline{u}_q, \underline{f}_q) & & \text{Presmoothing} \\
\underline{d}_q &:= \underline{f}_q - K_q \cdot \widehat{\underline{u}}_q & & \text{Defect calculation (new r.h.s.)} \\
\underline{d}_{q-1} &:= \mathfrak{I}_q^{q-1} \cdot \underline{d}_q & & \text{Restriction of defect} \\
\underline{w}_{q-1}^0 &:= 0 & & \text{Initial guess} \\
\underline{w}_{q-1} &:= \text{PMGM}^\gamma(K_{q-1}, \underline{w}_{q-1}^0, \underline{d}_{q-1}, q-1) & & \text{Solve defect system} \\
\underline{w}_q &:= \mathfrak{I}_{q-1}^q \cdot \underline{w}_{q-1} & & \text{Interpolation of correction} \\
\widetilde{\underline{u}}_q &:= \widehat{\underline{u}}_q + \underline{w}_q & & \text{Addition of correction} \\
\underline{u}_q &:= S_{\text{post}}^{\nu_{\text{post}}}(K_q, \widetilde{\underline{u}}_q, \underline{f}_q) & & \text{Postsmoothing}
\end{aligned}$

endif

Algorithm 7.2: Parallel multigrid: PMGM$(K_q, \underline{u}_q, \underline{f}_q, q)$.

The interested reader can find additional information on geometrical and algebraic multigrid methods and their parallelization in the survey paper [57].

Exercises

We want to solve the Laplace problem using the weak formulation with homogeneous Dirichlet boundary conditions on the discretization given in Fig. 5.4. Here we solve the resulting system of equations by means of a multigrid iteration.

E7.1. You will find a sequential version of the multigrid solver using ω-Jacobi iterations as a smoother and, for simplicity, as a coarse grid solver in the directory *Solution/mgseqc*. The following functions are added to the functions from the Exercises for Chapters 5 and 6.

AllocLevels(nlevels, nx, ny, pp, ierr)

allocates the memory for all levels and stores the pointers in variable **pp**, which is a structure **PointerStruct**. For details, see *mg.h*. The matrices and right-hand sides will be calculated in

IniLevels(xl, xr, yb, yt, nlevels, neigh, pp, ierr).
FreeLevels(nlevels, pp, ierr)

deallocates the memory at the end. Note that we need a different procedure for multigrid to apply the Dirichlet boundary conditions (BCs):

ApplyDirichletBC1(nx, ny, neigh, u, sk, id, ik, f).

The multigrid solver

MGMSolver(pp, crtl, level, neigh)

uses **crtl** as predefined structure **ControlStruct** to control the multigrid cycle (see *mg.h*). The same holds for each mg-iteration

MGM(pp, crtl, level, neigh).

The multigrid components smoothing, coarse grid solver, interpolation, and restriction are defined in

JacobiSmooth(nx, ny, sk, id, ik, f, u, dd, aux, maxiter),
JacobiSolve2(nx, ny, sk, id, ik, f, u, dd, aux),
Interpolate(nx, ny, w, nxc, nyc, wc, neigh),
Restrict(nx, ny, w, nxc, nyc, wc, neigh).

A vector $\underline{u}$ on level *level* can be saved in a file named *name* by calling

**SaveVector(name, pp->u[level], pp->nx[level], pp->ny[level],
u, nx, ny, xl, xr, yb, yt, ierr)**,

such that `gnuplot name` can be used to visualize that vector.

Your assignment, should you choose to accept it, is to implement a parallel version of the sequential multigrid code, i.e., Algorithm 7.2. These instructions will self-destruct in ten seconds.

E7.2. Replace the parallel ω-Jacobi solver on the coarse grid with the sequential ω-Jacobi solver on one process. This requires a gather of the right-hand side and a scatter of the solution after solving. Compare the run time of this parallel multigrid for solving the system of equations on the finest grid with the run time of the parallel multigrid from **E7.1**. Interpret the different run-time behavior.

Chapter 8

Problems Not Addressed in This Book

Prediction is difficult, especially of the future.
—Niels Bohr (1885–1962)

It is not the goal of this tutorial to compile a complete compendium on partial differential equations (PDEs), solvers, and parallelization. Each section of this tutorial can still be extended by many, many more methods as well as by theory covering other types of PDEs. Depending on the concrete problem the following topics, not treated in the book, may be of interest:

- Nonsymmetric problems, time-dependent PDEs, nonlinear PDEs, and coupled PDE problems [47, 103, 108].

- Error estimators and adaptive solvers [90, 109].

- Algorithms and programs that calculate the decomposition of nodes or elements to achieve a static load balancing, e.g., the programs Chaco [63], METIS [69], and JOSTLE [112]. See also [7, 74, 55] for basics and extensions of the techniques used.

- Dynamic load balancing [9, 17].

- The wide range of domain decomposition (DD) algorithms. A good overview of research activities in this field can be found in the proceedings of the Domain Decomposition Conferences (see [42, 18, 19, 43, 70, 87, 72, 44, 10, 78, 76, 20]) and in the collection [71]. Two pioneering monographs give a good introduction to DD methods from two different points of view: the algebraic one [97] and the analytic one [88].

- ....   (add a topic of your choice)

As already mentioned in the introduction to this tutorial, (PDE-based) parallel computational methods become more and more important in many sciences where they were not, or almost never, used 20 or even 10 years ago. This is especially true for the so-called life sciences such as computational biology and computational medicine (see, e.g., [35]) and the business sciences such as computational finance (see, e.g., [95]), where PDE models now play an important role.

Appendix

Internet Addresses

Software for parallelization

- The message-passing interface (MPI) homepage[1] should be used as a reference address for MPI. The MPI forum[2] contains the official MPI documents, especially the function calls in MPI;[3] see also another very good reference.[4]

- There exist several different implementations of the MPI standard. We use mainly the MPICH[5] implementation or the LAM[6] version.

- The Parallel Virtual Machine (PVM[7]) is an older parallel computing environment but it is still alive.

- The OpenMP[8] homepage contains the specifications of this parallelization approach.

Other software

- The Fast Fourier Transformation (FFT[9]) homepage should be visited for more information on that approach.

- Numerical software can be found in the Netlib.[10]

- Tools and benchmarks[11] for parallelization are also available.

---

[1] http://www-unix.mcs.anl.gov/mpi
[2] http://www.mpi-forum.org
[3] http://www.mpi-forum.org/docs/mpi-11-html/mpi-report.html
[4] http://www-unix.mcs.anl.gov/mpi/www
[5] http://www-unix.mcs.anl.gov/mpi/mpich
[6] http://www.lam-mpi.org
[7] http://www.csm.ornl.gov/pvm/pvm_home.html
[8] http://www.openmp.org
[9] http://www.fftw.org
[10] http://www.netlib.org
[11] http://www.nhse.org/ptlib

- METIS,[12] JOSTLE,[13] and CHACO[14] are well-known software packages for data distribution.

Parallel computers

- The Beowulf[15] Project allows parallel programming on LINUX clusters.

- One interesting project to construct some sort of super parallel computer is called Globus.[16]

- You can try to find your supercomputer in the TOP 500[17] list.

- The main vendors for parallel computers (as of 2002) are Cray,[18] Fujitsu,[19] HP,[20] IBM,[21] NEC,[22] SGI,[23] and SUN.[24]

People

- For more information contact the authors of this tutorial, Craig C. Douglas,[25] Gundolf Haase,[26] and Ulrich Langer.[27]

Further links

- A very nice overview of Parallel and High Performance Computing Resources is available.[28]

- The book by Pacheco[29] [84] is available on the internet.

- The porting of commercial software on parallel computers has been done in the EUROPORT[30] project.

---

[12]http://www-users.cs.umn.edu/~karypis/metis
[13]http://www.gre.ac.uk/~c.walshaw/jostle
[14]http://www.cs.sandia.gov/CRF/chac.html
[15]http://www.beowulf.org
[16]http://www.globus.org
[17]http://www.top500.org
[18]http://www.cray.com
[19]http://primepower.fujitsu.com/hpc/en/products-e/index-e.html
[20]http://www.hp.com/products1/servers/scalableservers/
[21]http://www.rs6000.ibm.com/hardware/largescale
[22]http://www.sw.nec.co.jp/hpc/sx-e
[23]http://www.sgi.com/origin/
[24]http://www.sun.com/servers/highend
[25]http://www.mgnet.org/~douglas
[26]http://www.numa.uni-linz.ac.at/Staff/haase/haase.html
[27]http://www.numa.uni-linz.ac.at/Staff/langer/langer.html
[28]http://www.jics.utk.edu/parallel.html
[29]http://www.usfca.edu/mpi
[30]http://www.gmd.de/SCAI/europort

- Find more information on multigrid[31] and domain decomposition[32] methods.

- One of the good sources for recent developments in scientific computing is the German Scientific Computing Page.[33]

---

[31]`http://www.mgnet.org`
[32]`http://www.ddm.org`
[33]`http://www.scicomp.uni-erlangen.de`

Bibliography

[1] R. A. ADAMS, *Sobolev Spaces*, Academic Press, San Francisco, 1975.

[2] O. AXELSSON, *Iterative Solution Methods*, Cambridge University Press, Cambridge, U.K., 1994.

[3] O. AXELSSON AND V. A. BARKER, *Finite Element Solution of Boundary Value Problems*, Academic Press, New York, 1984.

[4] I. BABUŠKA, *Courant element: Before and after*, in Finite Element Methods (Jyväskylä, 1993), Marcel Dekker, Inc., New York, 1994, pp. 37–51.

[5] R. E. BANK AND C. C. DOUGLAS, *Sharp estimates for multigrid rates of convergence with general smoothing and acceleration*, SIAM J. Numer. Anal., 22 (1985), pp. 617–633.

[6] R. E. BANK AND D. J. ROSE, *Some error estimates for the box method*, SIAM J. Numer. Anal., 24 (1987), pp. 777–787.

[7] S. T. BARNARD AND H. D. SIMON, *Fast multilevel implementation of recursive spectral bisection for partitioning unstructured problems*, Concurrency: Practice and Experience, 6 (1994), pp. 101–107.

[8] R. BARRETT, M. BERRY, T. F. CHAN, J. DEMMEL, J. DONATO, J. DONGARRA, V. EIJKHOUT, R. POZO, C. ROMINE, AND H. VAN DER VORST, *Templates for the Solution of Linear Systems: Building Blocks for Iterative Methods*, SIAM, Philadelphia, 1994. Available online at http://www.netlib.org/linalg/html_templates/report.html.

[9] P. BASTIAN, *Load balancing for adaptive multigrid methods*, SIAM J. Sci. Comput., 19 (1998), pp. 1303–1321.

[10] P. E. BJØRSTAD, M. ESPEDAL, AND D. KEYES, eds., *Domain Decomposition Methods in Sciences and Engineering*. Proceedings from the Ninth International Conference, June 1996, Bergen, Norway, John Wiley & Sons, New York, 1997.

[11] D. BRAESS, *Finite Elements: Theory, Fast Solvers and Applications in Solid Mechanics*, Cambridge University Press, Cambridge, U.K., 1997.

[12] J. H. BRAMBLE, *Multigrid Methods*, Pitman Research Notes in Mathematics Series, No. 294, Longman Scientific & Technical, Harlow, U.K., 1993.

[13] J. H. BRAMBLE, J. E. PASCIAK, AND A. H. SCHATZ, *The construction of preconditioners for elliptic problems by substructuring* I–IV, Math. Comput., 47 (1986), pp. 103–134; 49 (1987), pp. 1–16; 51 (1988), pp. 415–430; 53 (1989), pp. 1–24.

[14] A. BRANDT, *Multi-level adaptive solutions to boundary-value problems*, Math. Comput., 31 (1977), pp. 333–390.

[15] W. L. BRIGGS AND V. E. HENSON, *The DFT. An owner's Manual for the Discrete Fourier Transformation*, SIAM, Philadelphia, 1995.

[16] W. L. BRIGGS, V. E. HENSON, AND S. F. MCCORMICK, *A Multigrid Tutorial*, 2nd edition, SIAM, Philadelphia, 2000.

[17] A. CAGLAR, M. GRIEBEL, M. A. SCHWEITZER, AND G. ZUMBUSCH, *Dynamic load-balancing of hierarchical tree algorithms on a cluster of multiprocessor PCs and on the Cray T3E*, in Proceedings 14th Supercomputer Conference, Mannheim, H. W. Meuer, ed., Mateo, 1999.

[18] T. F. CHAN, R. GLOWINSKI, J. PÉRIAUX, AND O. B. WIDLUND, eds., *Second International Symposium on Domain Decomposition Methods for Partial Differential Equations*, Los Angeles, California, January 14-16, 1988, SIAM, Philadelphia, 1989.

[19] T. F. CHAN, R. GLOWINSKI, J. PÉRIAUX, AND O. B. WIDLUND, eds., *Third International Symposium on Domain Decomposition Methods for Partial Differential Equations*, Houston, March 20-22, 1989, SIAM, Philadelphia, 1990.

[20] T. F. CHAN, T. KAKO, H. KAWARADA, AND O. PIRONNEAU, eds., *Domain Decomposition Methods in Sciences and Engineering*, Proceedings of the 12th International Conference on Domain Decomposition, October 25-29, 1999, Chiba, Japan, DDM.org, Augsburg, 2001. Available electronically at http://www.ddm.org/DD12/index.html

[21] R. CHANDRA, R. MENON, L. DAGUM, D. KOHR, D. MAYDAN, AND J. MCDONALD, *Parallel Programming in OpenMP*, Morgan Kaufmann Publishers, San Francisco, 2000.

[22] P. CIARLET, *The Finite Element Method for Elliptic Problems*, Classics in Appl. Math. 40, SIAM, Philadelphia, 2002.

[23] J. W. COOLEY AND J. W. TUKEY, *An algorithm for the machine calculation of complex Fourier series*, Math. Comput., 19 (1965), pp. 297–301.

[24] R. COURANT, *Variational methods for the solution of problems of equilibrium and vibrations*, Bull. Amer. Math. Soc., 49 (1943), pp. 1–23.

[25] E. W. DIJKSTRA, *Cooperating sequential processes*, in Programming Languages, F. Genuys, ed., Academic Press, London, 1968, pp. 43–112.

[26] J. J. DONGARRA, I. S. DUFF, D. C. SORENSEN, AND H. A. VAN DER VORST, *Numerical Linear Algebra For High-Performance Computers*, vol. 7 of Software, Environments, and Tools, SIAM, Philadelphia, 1998.

[27] C. C. DOUGLAS, *Multi-grid algorithms with applications to elliptic boundary-value problems*, SIAM J. Numer. Anal., 21 (1984), pp. 236–254.

[28] C. C. DOUGLAS AND J. DOUGLAS, JR., *A unified convergence theory for abstract multigrid or multilevel algorithms, serial and parallel*, SIAM J. Numer. Anal., 30 (1993), pp. 136–158.

[29] C. C. DOUGLAS, J. DOUGLAS, AND D. E. FYFE, *A multigrid unified theory for non-nested grids and/or quadrature*, East-West J. Numer. Math., 2 (1994), pp. 285–294.

[30] J. DOUGLAS AND T. DUPONT, *Alternating-direction Galerkin methods on rectangles*, in Numerical Solution of Partial Differential Equations II, Academic Press, New York, 1971, pp. 133–214.

[31] J. DOUGLAS, R. B. KELLOGG, AND R. S. VARGA, *Alternating direction iteration methods for n space variables*, Math. Comput., 17 (1963), pp. 279–282.

[32] J. DOUGLAS AND D. W. PEACEMAN, *Numerical solution of two-dimensional heat flow problems*, Amer. Inst. Chem. Engrg. J., 1 (1955), pp. 505–512.

[33] M. DRYJA, *A finite element-capacitance matrix method for elliptic problems on regions partitioned into substructures*, Numer. Math., 34 (1984), pp. 153–168.

[34] D. J. EVANS, G. R. JOUBERT, AND H. LIDELL, eds., *Parallel Computing '91*, Proceedings of the International Conference on Parallel Computing, September, 1991, London, U.K., North-Holland, Amsterdam, 1992.

[35] C. FALL, E. MARLAND, J. WAGNER, AND J. TYSON, *Computational Cell Biology*, Springer, New York, 2002.

[36] J. H. FERZIGER AND M. PERIC, *Computational Methods for Fluid Dynamics*, Springer-Verlag, Berlin, 1999.

[37] M. FLYNN, *Very high-speed computing systems*, Proc. IEEE, 54 (1966), pp. 1901–1909.

[38] I. FOSTER, *Designing and Building Parallel Programs*, Addison-Wesley, Reading, MA, 1994.

[39] P. J. FREY AND P. L. GEORGE, *Mesh Generation: Applications to Finite Elements*, Hermes Science, Paris, Oxford, 2000.

[40] K. GALLIVAN, M. HEATH, E. NG, B. PEYTON, R. PLEMMONS, J. ORTEGA, C. ROMINE, A. SAMEH, AND R. VOIGT, *Parallel Algorithms for Matrix Computations*, SIAM, Philadelphia, 1990.

[41] F. GENUYS, ed., *Programming Languages*, Academic Press, London, 1968.

[42] R. GLOWINSKI, G. H. GOLUB, G. A. MEURANT, AND J. PÉRIAUX, eds., *First International Symposium on Domain Decomposition Methods for Partial Differential Equations*, January 1987, Paris. SIAM, Philadelphia, 1988.

[43] R. GLOWINSKI, Y. A. KUZNETSOV, G. MEURANT, J. PÉRIAUX, AND O. B. WIDLUND, eds., *Fourth International Symposium on Domain Decomposition Methods for Partial Differential Equations*, May 1990, Moscow. SIAM, Philadelphia, 1991.

[44] R. GLOWINSKI, J. PÉRIAUX, AND Z. SHI, eds., *Domain Decomposition Methods in Sciences and Engineering*, Proceedings of Eighth International Conference, Beijing, P.R. China. John Wiley & Sons, New York, 1997.

[45] G. GOLUB AND J. M. ORTEGA, *Scientific Computing: An Introduction with Parallel Computing*, Academic Press, New York, 1993.

[46] G. H. GOLUB AND C. F. VAN LOAN, *Matrix Computations*, 3rd edition, Johns Hopkins University Press, Baltimore, MD, 1996.

[47] M. GRIEBEL, T. DORNSEIFER, AND T. NEUNHOEFFER, eds., *Numerical Simulation in Fluid Dynamics: A Practical Introduction*, vol. 3 of SIAM Monographs on Mathematical Modeling and Computation, SIAM, Philadelphia, 1998.

[48] U. GROH, *Local Realization of Vector Operations on Parallel Computers*, Preprint SPC 94-2, TU Chemnitz, 1994. In German. Available online at http://www.tu-chemnitz.de/ftp-home/pub/Local/mathematik/SPC/spc94_2.ps.gz

[49] W. GROPP, E. LUSK, AND A. SKJELLUM, *Using MPI*, MIT Press, Cambridge, MA, London, 1994.

[50] C. GROSSMANN AND H.-G. ROOS, *Numerik partieller Differentialgleichungen*, B.G. Teubner, Stuttgart, 1992. In German.

[51] I. GUSTAFSSON, *A class of first order factorization methods*, BIT, 18 (1978), pp. 142–156.

[52] G. HAASE, *Parallel incomplete Cholesky preconditioners based on the non-overlapping data distribution*, Parallel Computing, 24 (1998), pp. 1685–1703.

[53] G. HAASE, *Parallelisierung numerischer Algorithmen für partielle Differentialgleichungen*, Teubner, Leipzig, 1999.

[54] G. HAASE, *A parallel AMG for overlapping and non-overlapping domain decomposition*, Electronic Trans. Numer. Anal. (ETNA), 10 (2000), pp. 41–55.

[55] G. HAASE AND M. KUHN, *Preprocessing in 2D FE-BE domain decomposition methods*, Comput. Visualization Sci., 2 (1999), pp. 25–35.

[56] G. HAASE, M. KUHN, AND S. REITZINGER, *Parallel algebraic multigrid methods on distributed memory computers*, SIAM J. Sci. Comput., 24 (2002), pp. 410–427.

[57] G. HAASE AND U. LANGER, *Multigrid methods: From geometrical to algebraic versions*, in Modern Methods in Scientific Computing and Applications, Kluwer Academic Press, Dordrecht, 2002. Vol. 75 in the NATO Science Ser. II, Chapter X.

[58] W. HACKBUSCH, *Multi-Grid Methods and Applications*, vol. 4 of Computational Mathematics, Springer-Verlag, Berlin, 1985.

[59] W. HACKBUSCH, *On first and second order box schemes*, Computing, 41 (1989), pp. 277–296.

[60] W. HACKBUSCH, *Iterative Solution of Large Sparse Systems*, Springer-Verlag, Berlin, 1994.

[61] D. W. HEERMANN AND A. N. BURKITT, *Parallel Algorithms in Computational Science*, vol. 24 of Information Sciences, Springer-Verlag, Berlin, 1991.

[62] B. HEINRICH, *Finite Difference Methods on Irregular Networks*, vol. 82 of International Series of Numerical Mathematics, Birkhäuser-Verlag, Basel, 1986.

[63] B. HENDRICKSON AND R. LELAND, *The Chaco User's Guide – Version* 2.0, Technical Report SAND94-2692, Sandia National Laboratories, Albuquerque, NM, 1992.

[64] K. HWANG, *Advanced Computer Architecture: Parallelism, Scalability, Programmabilty*, McGraw-Hill Computer Engineering Series, McGraw-Hill, New York, 1993.

[65] E. ISSACSON AND H. KELLER, *Analysis of Numerical Methods*, Dover Publications, Mineola, NY, 1994. Reprint of a 1966 text.

[66] M. JUNG, *On the parallelization of multi-grid methods using a non-overlapping domain decomposition data structure*, Appl. Numer. Math., 23 (1997), pp. 119–137.

[67] M. JUNG AND U. LANGER, *Application of multilevel methods to practical problems*, Surveys Math. Industry, 1 (1991), pp. 217–257.

[68] M. JUNG, U. LANGER, A. MEYER, W. QUECK, AND M. SCHNEIDER, *Multigrid preconditioners and their applications*, in Third Multigrid Seminar, Biesenthal 1988, G. Telschow, ed., Berlin, 1989, Karl–Weierstrass–Institut, pp. 11–52. Report R–MATH–03/89.

[69] G. KARYPIS AND V. KUMAR, *METIS: Unstructured Graph Partitioning and Sparse Matrix Ordering System—version* 2.0, Technical Report, Department of Computer Science, University of Minnesota, 1995.

[70] D. E. KEYES, T. F. CHAN, G. MEURANT, J. S. SCROGGS, AND R. G. VOIGT, eds., *Fifth International Symposium on Domain Decomposition Methods for Partial Differential Equations*, May 1991, Norfolk. SIAM, Philadelphia, 1992.

[71] D. E. KEYES, Y. SAAD, AND D. G. TRUHLAR, eds., *Domain-Based Parallelism and Problem Decomposition Methods in Computational Science and Engineering*, SIAM, Philadelphia, 1995.

[72] D. E. KEYES AND J. XU, eds., *Domain Decomposition Methods in Scientific and Engineering Computing*, Proceedings of the Seventh International Conference on Domain Decomposition, October 1993, Pennsylvania State University. AMS, Providence, RI, 1994.

[73] N. KIKUCHI, *Finite Element Methods in Mechanics*, Cambridge University Press, Cambridge, U.K., 1986.

[74] R. KOPPLER, G. KURKA, AND J. VOLKERT, *Exdasy—a user-friendly and extendable data distribution system*, in Euro-Par '97 Parallel Processing, C. Lengauer, M. Griebel, and S. Gorlatsch, eds., Lecture Notes Comput. Sci., 1300, Springer-Verlag, Berlin, 1997, pp. 118–127.

[75] V. KUMAR, A. GRAMA, A. GUPTA, AND G. KARYPIS, *Introduction to Parallel Computing: Design and Analysis of Algorithms*, Benjamin/Cummings Publishing Company, Redwood City, CA, 1994.

[76] C.-H. LAI, P. E. BJØRSTAD, M. CROSS, AND O. B. WIDLUND, eds., *Eleventh International Conference on Domain Decomposition Methods*, (Greenwich, England, July 20-24, 1998), DDM.org, Augsburg, 1999. Available electronically at http://www.ddm.org/DD11/index.html

[77] K. H. LAW, *A parallel finite element solution method*, Comput. Struct., 23 (1989), pp. 845–858.

[78] J. L. MANDEL, ed., *Domain Decomposition Methods 10—The Tenth International Conference on Domain Decomposition Methods*, August 1997, Boulder, CO. AMS, Providence, RI, 1998.

[79] J. MEIJERINK AND H. VAN DER VORST, *An iterative solution for linear systems of which the coefficient matrix is a symmetric m-matrix*, Math. Comput., 31 (1977), pp. 148–162.

[80] G. MEURANT, *Computer Solution of Large Linear Systems*, North-Holland, Amsterdam, 1999.

[81] A. MEYER, *On the $O(h^{-1})$-Property of MAF*, Preprint 45, TU Karl-Marx-Stadt, Chemnitz, Germany, 1987.

[82] A. MEYER, *A parallel preconditioned conjugate gradient method using domain decomposition and inexact solvers on each subdomain*, Computing, 45 (1990), pp. 217–234.

[83] R. MILLER AND Q. F. STOUT, *Parallel Algorithms for Regular Architectures: Meshes and Pyramid*, MIT Press, Cambridge, MA, 1996.

[84] P. PACHECO, *Parallel Programming with MPI*, Morgan Kaufmann Publishers, San Francisco, 1997.

[85] D. W. PEACEMAN AND H. H. RACHFORD, *The numerical solution of parabolic and elliptic differential equations*, J. Soc. Indust. Appl. Math., 3 (1955), pp. 28–41.

[86] C. M. PEARCY, *On convergence of alternating direction procedures*, Numer. Math., 4 (1962), pp. 172–176.

[87] A. QUARTERONI, J. PERIAUX, Y. KUZNETSOV, AND O. B. WIDLUND, eds., *Sixth International Conference on Domain Decomposition Methods in Science and Engineering*, AMS, Providence, RI, 1994.

[88] A. QUARTERONI AND A. VALI, *Domain Decomposition Methods for Partial Differential Equations*, Oxford Science Publications, Oxford, UK, 1999.

[89] S. H. ROOSTA, *Parallel Programming and Parallel Algorithms: Theory and Computation*, Springer-Verlag, New York, 2000.

[90] U. RÜDE, *Mathematical and Computational Techniques for Multilevel Adaptive Methods*, vol. 13 of Frontiers in Applied Mathematics, SIAM, Philadelphia, 1993.

[91] J. W. RUGE AND K. STÜBEN, *Algebraic multigrid*, in Multigrid Methods, S. F. McCormick, ed., vol. 5 of Frontiers in Applied Mathematics, SIAM, Philadelphia, 1987, pp. 73–130.

[92] Y. SAAD, *Iterative Methods for Sparse Linear Systems*, PWS Publishing Company, Boston, 1995.

[93] Y. SAAD AND M. H. SCHULTZ, *GMRES: A generalized minimal residual algorithm for solving nonsymmetric linear systems*, SIAM J. Sci. Stat. Comput., 7 (1986), pp. 856–869.

[94] A. A. SAMARSKI AND E. S. NIKOLAEV, *Numerical Methods for Grid Equations*, vol. I/II, Birkhäuser-Verlag, Basel, Boston, Berlin, 1989.

[95] R. SEYDEL, *Tools for Computational Finance*, Springer-Verlag, Heidelberg, 2002.

[96] SGI, *Origin Servers*, Technical Report, Silicon Graphics Computer Systems, Mountain View, CA, 1997.

[97] B. F. SMITH, P. E. BJØRSTAD, AND W. GROPP, *Domain Decomposition: Parallel Multilevel Methods for Elliptic Partial Differential Equations*, Cambridge University Press, Cambridge, U.K., 1996.

[98] G. STRANG AND G. J. FIX, *An Analysis of the Finite Element Method*, Prentice–Hall, New York, 1973.

[99] G. THOMAS AND C. W. UEBERHUBER, *Visualization of Scientific Parallel Programs*, Springer-Verlag, Berlin, 1994.

[100] J. F. THOMPSON, B. S. BHARAT, AND N. P. WEATHERRILL, *Handbook of Grid Generation*, CRC Press, Philadelphia, 1998.

[101] C. H. TONG, T. F. CHAN, AND C. C. JAY KUO, *Multilevel filtering preconditioners: Extensions to more general elliptic problems*, SIAM J. Sci. Stat. Comput., 13 (1992), pp. 227–242.

[102] B. H. V. TOPPING AND A. I. KHAN, *Parallel Finite Element Computations*, Saxe-Coburg Publications, Edinburgh, 1996.

[103] A. TVAITO AND R. WINTHER, *Introduction to Partial Differential Equations, A Computational Approach*, Springer-Verlag, New York, 1998.

[104] E. VAN DE VELDE, *Concurrent Scientific Computing*, Texts in Applied Mathematics, Springer-Verlag, New York, 1994.

[105] H. A. VAN DER VORST, *High performance preconditioning*, SIAM J. Sci. Stat. Comput., 10 (1989), pp. 1174–1185.

[106] C. VAN LOAN, *Computational Framework for the Fast Fourier Transformation*, vol. 10 of Frontiers in Applied Mathematics, SIAM, Philadelphia, 1992.

[107] U. VAN RIENEN, *Numerical Methods in Computational Electrodynamics*, Springer-Verlag, Berlin, Heidelberg, 2000.

[108] S. VANDEWALLE, *Parallel Multigrid Waveform Relaxation for Parabolic Problems*, Teubner Skripten zur Numerik, Teubner, 1993.

[109] R. VERFÜRTH, *A Review of a Posteriori Error Estimation and Adaptive Mesh-Refinement Techniques*, Wiley-Teubner, Chichester, 1996.

[110] E. L. WACHSPRESS, *The ADI Model Problem*, Wachspress, Windsor, CA, 1995.

[111] C. WALSHAW AND M. CROSS, *Mesh partitioning: A multilevel balancing and refinement algorithm*, SIAM J. Sci. Comput., 22 (2000), pp. 63–80.

[112] C. WALSHAW, M. CROSS, S. P. JOHNSON, AND M. G. EVERETT, *JOSTLE: Partitioning of unstructured meshes for massively parallel machines*, in Proceedings of the Parallel CFD'94, Kyoto, N. Satofuka, ed., Elsevier, Amsterdam, 1994, pp. 273–280.

[113] W. L. WENDLAND, ed., *Boundary Element Topics*, Proceedings of the Final Conference of the Priority Research Programme Boundary Element Methods 1989–1995 of the German Research Foundation, Springer-Verlag, Berlin, Heidelberg, 1998.

[114] P. WESSELING, *An Introduction to Multigrid Methods*, Cos Cob, CT, 2001. Reprint of the 1992 edition. Available online at http://www.mgnet.org/mgnet-books.html

[115] E. ZEIDLER, *Applied Functional Analysis: Applications to Mathematical Physics*, vol. 108 of Applied Mathematical Sciences, Teubner, Stuttgart, 1995.

[116] O. C. ZIENKIEWICZ, *The Finite Element Method in Engineering Science*, McGraw Hill, London, 1971.

Index